VOICE OVER LTE

VOICE OVER LTE

VoLTE

Miikka Poikselkä, Harri Holma, Jukka Hongisto,
Juha Kallio and Antti Toskala
Nokia Siemens Networks, Finland

WILEY
A John Wiley & Sons, Ltd., Publication

This edition first published 2012

Registered office
John Wiley & Sons Ltd, The Atrium, Southern Gate, Chichester, West Sussex, PO19 8SQ, United Kingdom

For details of our global editorial offices, for customer services and for information about how to apply for permission to reuse the copyright material in this book please see our website at www.wiley.com.

Library of Congress Cataloging-in-Publication Data

Poikselkä, Miikka.
Voice over LTE (VoLTE) / Miikka Poikselkä, Harri Holma, Jukka Hongisto, Juha Kallio and Antti Toskala.
pages cm
Includes bibliographical references and index.
ISBN 978-1-119-95168-1
1. Long-Term Evolution (Telecommunications) 2. Internet telephony. I. Holma, Harri, 1970- II. Hongisto, Jukka. III. Kallio, Juha. IV. Toskala, Antti. V. Title.
TK5103.48325.P65 2012
621.3845′6–dc23

2011047199

A catalogue record for this book is available from the British Library.

Print ISBN: 9781119951681

Set in 10/12 Times by Laserwords Private Limited, Chennai, India
Printed and bound in Great Britain by CPI Group (UK) Ltd, Croydon, CR0 4YY

Contents

Preface

Voice communication has become mobile in a massive way and the mobile is the preferred way for voice communication for more than six billion subscribers. The introduction of high speed packet access (HSPA) also allows a large set of data services from smartphones, tablets and laptops to go mobile. The data volumes in mobile networks greatly exceed the voice volumes – from the traffic point of view, networks have already turned from voice dominated to data dominated. The next generation mobile radio system, called long term evolution (LTE), started commercially in 2009 and is designed to boost further the data rates and capacities. LTE radio is primarily optimised for high capacity data but can also support voice efficiently. During 2012, smartphones with the capability to provide a voice service using the LTE network are becoming available. This book describes how the voice service is supported with LTE capable terminals. The voice support in LTE is not as trivial as in second generation (2G) GSM and third generation (3G) WCDMA solutions where circuit switched (CS) voice was used, since LTE is designed only for packet switched (PS) connections. The voice service in LTE uses the voice over Internet protocol (VoIP) – called VoLTE (voice over LTE) – together with the IP multimedia system (IMS). There are also alternative solutions for supporting voice in the initial phase where the voice service runs on legacy 2G/3G networks while only data is carried on LTE. The voice solutions, architectures and required functionalities are described in this book.

The book is structured as follows. Chapter 1 gives the background and Chapter 2 describes the operator VoLTE deployment strategies. The system architecture is presented in Chapter 3 and VoLTE functionalities in Chapter 4. End to end signalling is illustrated in Chapter 5 and IMS centralized services in Chapter 6. The VoLTE radio performance is presented in Chapter 7 and the VoIP over HSPA networks in Chapter 8.

LTE will access a very large global market – not only GSM/WCDMA operators, but also CDMA and WiMAX operators and potentially also fixed network service providers. The large potential market can attract a large number of companies to the market place pushing the economics of scale, which enables wide scale LTE adoption with lower cost. The voice solution is a key part of the system design – every single LTE capable smartphone needs to have a good quality voice solution. This book is particularly designed for chip set and mobile vendors, network vendors, network operators, technology managers and regulators who would like to get a deeper understanding of voice over LTE.

Acknowledgements

We would like to thank the following colleagues for their valuable comments: Jari Välimäki, Peter Leis, Curt Wong and Martin Öttl.

The authors appreciate the fast and smooth editing process provided by John Wiley & Sons Ltd and especially Susan Barclay, Sandra Grayson, Sarah Tilley, Sophia Travis and Mark Hammond.

We are grateful to our families for their patience during the late night and weekend editing sessions.

The authors welcome any comments and suggestions for improvements or changes that could be used to improve future editions of this book. The feedback is welcomed to the authors' email addresses - miikka.poikselka@nsn.com, harri.holma@nsn.com, jukka.hongisto@nsn.com, juha.kallio@nsn.com and antti.toskala@nsn.com.

List of Abbreviations

1×RTT	Single carrier radio transmission technology
3GPP	Third generation partnership project
3GPP2	Third generation partnership project 2
AAA	Authentication, authorization and accounting
AAL	ATM adaptation layer
AAR	AA request
ACA	Accounting answer
ACK	Acknowledgement
ACR	Accounting request
AGCF	Access gateway control function
A-GNSS	Assisted global navigation satellite system
AGW	Access gateway
AKA	Authentication and key agreement
ALG	Application level gateway
AMR	Adaptive multi-rate
AMR-WB	Adaptive multi-rate wideband
AOC	Advice of charge
AP	Authentication proxy
APN	Access point name
APN-AMBR	Access point name–aggregate maximum bit rate
ARP	Allocation and retention priority
AS	Application server
ASA	Abort session answer
ASR	Abort session request
ATCF	Access transfer control function
ATGW	Access transfer gateway
ATM	Asynchronous transfer mode
AUC	Authentication centre
AUID	Application usage identification
AUTN	Authentication token

AVC	Advanced video coding
AVP	Attribute value pair; audio visual profile
AVPF	Audio visual profile with feedback
B2BUA	Back to back user agent
BGCF	Breakout gateway control function
BICC	Bearer independent call control
BS	Billing system
BSC	Base station controller
BSS	Base station system
BTS	Base transceiver station
CAT	Customized alerting tone
CB	Communication barring
CCBS	Completion of communications to busy subscriber
CCF	Charging collection function
CCNR	Completion of communications on no reply
CD	Communication deflection
CDF	Charging data function
CDMA	Code division multiple access
CDR	Charging data record
CLI	Calling line identification
CN	Core network
CSFB	Circuit switched fallback
DCH	Dedicated traffic channel
PCH	Paging channel
CFB	Communication forwarding busy
CFNL	Communication forwarding on not logged in
CFNR	Communication forwarding no reply
CFNRc	Communication diversion on mobile subscriber not reachable
CFU	Communication forwarding unconditional
CGF	Charging gateway function
CGI	Cell global identification
CK	Ciphering key
CLIP	Calling line identification presentation
C-MSISDN	Correlation mobile subscriber international ISDN number
CONF	Conference calling
CPC	Continuous packet connectivity
CPE	Customer premise equipment
CRS	Customized ringing signal
CS	Circuit switched
CSCF	Call session control function
CSFB	Circuit switched fallback
CSMO	Circuit switched mobile originating
CTF	Charging trigger function

CUG	Closed user group
CW	Communication waiting
DCH	Dedicated channel
DFCA	Dynamic frequency and channel allocation
DL	Downlink
DNS	Domain name system
DOCSIS	Data over cable service interface specification
DPCCH	Dedicated physical control channel
DPDCH	Dedicated physical data channel
DRB	Data radio bearer
DRX	Discontinuous reception
DTAP	Direct transfer application part
DTM	Dual transfer mode
DTMF	Dual-tone Multi-frequency
DTX	Discontinuous transmission
EATF	Emergency access transfer function
ECF	Event charging function
ECM	Evolved packet system connection management
E-CSCF	Emergency CSCF
ECT	Explicit communication transfer
E-DCH	Enhanced dedicated channel
EDGE	Enhanced data rates for global evolution
EFR	Enhanced full rate
eHRPD	Evolved high rate packet data
EMM	Evolved packet system mobility management
eMSS	Enhanced mobile switching centre server
eNB	Evolved nodeB
eNodeB	Evolved nodeB
EPC	Evolved packet core
ePDG	Evolved packet data gateway
EPS	Evolved packet system
E-SMLC	Evolved serving mobile location centre
E-STN-SR	Emergency session transfer number for SRVCC
ETSI	European Telecommunications Standards Institute
E-UTRAN	Evolved universal mobile telecommunications system terrestrial radio access network
EV-DO	Evolution data only
FA	Flexible alerting
FGI	Feature group indicator
FoR	Follow on request
FTP	File transfer protocol
GAA	Generic authentication architecture
GBR	Guaranteed bit rate
GERAN	Global system for mobile communications/EDGE radio access network

GGSN	Gateway general packet radio service support node
GMLC	Gateway mobile location centre
GPRS	General packet radio service
GPS	Global positioning system
GRUU	Globally routable user agent URI
GSM	Global system for mobile communications
GSMA	GSM association
GTP	GPRS tunnelling protocol
GTP-C	GPRS tunnelling protocol control plane
GTP-U	GPRS tunnelling protocol user plane
GWF	Gateway function
HARQ	Hybrid automatic repeat request
HLR	Home location register
HOLD	Communication hold
HPLMN	Home public land mobile network
HRPD	High rate packet data
HSDPA	High speed downlink packet access
HS-DSCH	High speed downlink shared channel
HSGW	HRPD serving gateway
HSPA	High speed packet access
HSS	Home subscriber server
HSUPA	High speed uplink packet access
HTTP	Hyper text transfer protocol
IAM	Initial address message
IBCF	Interconnection border control function
ICID	Internet protocol multimedia subsystem charging identifier
ICS	Internet protocol multimedia subsystem centralised services
I-CSCF	Interrogating CSCF
ICSI	Internet protocol multimedia subsystem communication service identification
IETF	Internet engineering task force
iFC	Initial filter criteria
IK	Integrity key
IM	Instant messaging
IMEI	International mobile equipment identity
IMPI	IMS private user identity
IMPU	Internet protocol multimedia subsystem public user identity
IMS	Internet protocol multimedia subsystem
IMSI	International mobile subscriber identifier
IMS-MGW	IMS media gateway
IM-SSF	Internet protocol multimedia service switching function

IMTC	International multimedia teleconferencing consortium
IOI	Inter-operator identifier
IP	Internet protocol
IP-CAN	IP connectivity access network
IP-PBX	IP private branch exchange
IPsec	Internet protocol security
IP-SM-GW	IP short message gateway
IPv4	Internet protocol version 4
IPv6	Internet protocol version 6
ISC	IMS service control
ISDN	Integrated services digital network
ISIM	IP multimedia services identity module
ISR	Idle mode signaling reduction
ISUP	ISDN user part
LAI	Location area identity
LCS	Location control services
LIA	Location info answer
LIR	Location info request
LRF	Location retrieval function
LTE	Long term evolution
MAA	Multimedia–multimedia answer
MAC	Medium access control
MaM	Mid-call assisted mobile switching centre server
MAP	Mobile application part
MAR	Multimedia authorisation request
MBR	Maximum bit rate
MCC	Mobile country code
MCID	Malicious communication identification
ME	Mobile equipment
MGCF	Media gateway control function
MGW	Media gateway function
MIME	Multipurpose Internet mail extension
MI-USSD	Mobile initiated unstructured supplementary service data
MME	Mobility management entity
MMS	Multimedia messaging service
MMTel	Multimedia telephony communication service
MNC	Mobile network code
MO-LR	Mobile originating location request
MRB	Media resource broker
MRFC	Multimedia resource function controller
MRFP	Media resource function processor
MSC	Mobile switching centre
MSIN	Mobile subscriber identification number
MSISDN	Mobile subscriber ISDN

MSRN	Mobile station roaming number
MSRP	Message session relay protocol
MSS	MSC server
MT-LR	Mobile terminating location request
MTP	Message transfer part
MTRF	Mobile terminating roaming forwarding
MTRR	Mobile terminating roaming retry
MTU	Maximum transmission unit
MWI	Message waiting indication
NACC	Network assisted cell change
NAI	Network access identifier
NAS	Network access stratum
NAT	Network address translator
NB	Narrowband
NDS	Network domain security
NGMN	Next generation mobile networks alliance
NGN	Next generation network
NI-USSD	Network initiated unstructured supplementary service data
NW	Network
OCS	Online charging system
OFDMA	Orthogonal frequency division multiple access
OIP	Originating identification presentation
OIR	Originating identification restriction
OMA	Open mobile alliance
OSA	Open service architecture
OTDOA	Observed time difference of arrival
PCC	Policy and charging control
PCEF	Policy enforcement function
PCRF	Policy and charging rules function
P-CSCF	Proxy-CSCF
PDCCH	Physical downlink control channel
PDCP	Packet data convergence protocol
PDN	Packet data network
PDP	Packet data protocol
PDSCH	Physical downlink shared channel
P-GW	PDN gateway
PHY	Physical layer
PLMN	Public land mobile network
PNM	Personal network management
PoC	Push to talk over cellular
POTS	Plain old telephone service
PRACK	Provisional response acknowledgement
PRN	Provide roaming number
PS	Packet switched; presence server
PS HO	Packet switched handover

PSAP	Public safety answering point
PSTN	Public switched telephone network
QCI	QoS class identifier
QoS	Quality of service
RAA	Re-authorisation request
RAB	Radio access bearer
RAN	Radio access network
RAND	Random challenge
RAR	Re-authorisation request
RAT	Radio access technology
RCS	Rich communication suite
RDF	Routing determination function
RES	Response
RF	Radio frequency
RFC	Requests for comments
RLC	Radio link control
RNC	Radio network controller
RoHC	Robust header compression
RR	Receiver reports
RRC	Radio resource control
RSH	Resume call handling
RSRP	Reference symbol received power
RSRQ	Reference signal received quality
RTCP	Real-time transport protocol control protocol
RTP	Real-time transport protocol
RTP/AVP	RTP audio and video profile
S1AP	S1 application protocol
SA	Security association
SAA	Server assignment answer
SAE	System architecture evolution
SAI	Service area identifier
SAR	Server assignment request
SCC AS	Service centralisation and continuity application server
SCCP	Signalling connection control part
SC-FDMA	Single carrier FDMA
S-CSCF	Serving CSCF
SCTP	Stream control transmission protocol
SDP	Session description protocol
SEG	Security gateway
SGsAP	SGs application part
SGSN	Serving GPRS support node
S-GW	Serving gateway
SIB	System information block
SIM	Subscriber identity module
SIP	Session initiation protocol

SLF	Subscription locator function
SM	Short message
SMS	Short message service
SMSC	Short message service centre
SPR	Subscription profile repository
SPS	Semi persistent scheduling
SPT	Service point trigger
SQN	Sequence number
SR-VCC	Single radio voice call continuity
SRVCC	Single radio voice call continuity
SUPL	Secure user plane location
SV-DO	Simultaneous voice and data only
SV-LTE	Simultaneous voice and LTE
TA	Tracking area
TAI	Tracking area identity
TAS	Telephony application server
TCP	Transmission control protocol
TFT	Traffic flow template
THIG	Topology hiding inter-network gateway
TIP	Terminating identification presentation
TIR	Terminating identification restriction
TISPAN	Telecommunications and Internet converged services and protocols for advanced networking
TLS	Transport layer security protocol
TLV	Tag length value
TMR	Transport medium requirement
TMSI	Temporary mobile subscriber identity
TrGW	Transition gateway
TTI	Transmission time interval
UA	User agent
UDP	User datagram protocol
UE	User equipment
UE-AMBR	UE aggregate maximum bit rate
UICC	Universal integrated circuit card
UL	Uplink
UM	Unacknowldeged mode
UMTS	Universal mobile telecommunications system
URA	UTRAN registration area
URI	Uniform resource identifier
URL	Universal resource locator
USI	User service information
USIM	Universal subscriber identity module
USSD	Unstructured supplementary service data
USSI	Unstructured supplementary services data simulation in IMS
UTDOA	Uplink time difference of arrival

UTRAN	UMTS terrestrial radio access network
VLR	Visitor location register
VMSC	Visited MSC
VoIP	Voice over IP
VoLGA	Voice over LTE via generic access
VoLTE	Voice over LTE
VPLMN	Visited PLMN
WCDMA	Wideband code division multiple access
WLAN	Wireless local area network
XCAP	XML configuration access protocol
XDM	XML document management
XDMS	XML document management server
xDSL	Digital subscriber line
XML	Extensible markup language
XRES	Expected response

1

Background

At the end of 2004 the third generation partnership project (3GPP) Standardisation Forum started to evaluate a new radio technology as a successor for wideband code division multiple access (WCDMA). This work was called long term evolution (LTE) and is nowadays the radio interface name used in most official publications. Inside 3GPP the newly developed radio access network is called the evolved UMTS terrestrial radio access network (E-UTRAN) to indicate the path from the global system for mobile communications (GSM)/Enhanced data rates for global evolution (EDGE) radio access network (GERAN) via the GSM/general packet radio service (GPRS)/EDGE to UTRAN [WCDMA/high-speed packet access (HSPA)] and finally to E-UTRAN (LTE). In parallel to the work on a new radio interface 3GPP initiated a study to evolve the 2G/3G packet core network (known as the GPRS core) in order to cope with the new demands of LTE. This core network study was called system architecture evolution (SAE) and it was documented in the 3GPP technical report (3GPP TR 23.882). The final outcome of this work was a new packet core design in Release 8 documented in (3GPP TS 23.401) and (3GPP TS 23.402), called the evolved packet core (EPC). 3GPP Release 8 was officially completed in March 2009 and the world's first commercial LTE network was opened in December 2009 by TeliaSonera.

3GPP Release 8 introduces major advances in mobile networks. For the subscriber, it means higher access rates and lower latency on the connection, while for the mobile communication service provider, LTE radio technology provides lower cost per transmitted bit thanks to more efficient use of radio network resources and delivers excellent voice spectral efficiency, as described in Chapter 7. The technology also offers more flexibility in frequency allocation, thanks to the ability to operate LTE networks across a very wide spectrum of frequencies. LTE also minimises the power consumption of terminals that are used 'always on'. 3GPP Release 8 also introduces major advances in the core network that improve service quality and networking efficiency, leading to a better end user experience. GPRS technology has already introduced the always on concept for subscriber connectivity and 3GPP Release 8 mandates this ability, with at least one default bearer being always available for all subscribers. This allows fast access to services as well as network initiated services such as terminating voice calls and push e-mail. The connection setup time for person to person communication is also minimised with always on bearers.

Voice over LTE: VoLTE, First Edition. Miikka Poikselkä, Harri Holma, Jukka Hongisto, Juha Kallio and Antti Toskala.

But considering the fact that LTE is an all-internet protocol (IP) technology we can get to the conclusion that the voice service will have to be delivered in a different way as circuit switched voice will not be possible. So there is a need for a voice solution on top of LTE. Voice in this IP world, would be implemented as voice over Internet protocol (VoIP). The 3GPP-specified way to support VoIP is the IP multimedia subsystem (IMS). It is an access-independent and standard-based IP connectivity and service control architecture that enables various types of multimedia services to end users using common Internet-based protocols. 3GPP has worked with the IMS since 2000 and there exist thousands of pages, in different specifications, that cover IMS related functionalities. In the meantime, a sophisticated architecture and feature set has been developed. Moreover, 3GPP has specified multiple, different ways to complete single functions (e.g. authentication, session setup, supplementary service execution, bearer setup) which increases complexity of the IMS.

While 3GPP has specified all of the 'ingredients' needed to implement IMS-based voice over long term evolution (VoLTE) – such as session initiation protocol (SIP) registration, signalling compression, call set up and supplementary services – it has left it up to communication service providers and vendors to decide which of the numerous alternative implementation options to use. This is frankly a recipe for a chaotic and fragmented rollout of IMS-based VoLTE since there is no way to guarantee that different industry players will opt for the same 'ingredients' that their competitors' choose for their own implementations. It goes without saying it was not a model for success.

Unsurprisingly, in the absence of a clear-cut approach to VoLTE, alternatives emerged, most notably 3GPP specified circuit switched fallback (CSFB), in which an communication service provider uses its legacy 2G/3G network to handle voice calls. In this scenario, when an LTE terminal initiates a voice call or receives one from the legacy circuit-switched network, it downgrades any ongoing LTE data session to 3G or HSPA speeds for the duration of the voice call. If the voice call 'falls back' to a 2G network, the LTE data session will likely be suspended altogether, as 2G data speeds are not sufficient for broadband data applications. In either case, the impact on customer experience can be obvious.

Another emerged alternative for IMS-based VoLTE was VoLTE via generic access (VoLGA) promoted by the VOLGA Forum.

All major network and handset vendors compete aggressively for the biggest possible slice of network communication service providers' business. But at the same time, their business is an interconnected business, where equipment interoperability, especially between handset and network is the key to ensuring that they can all play together. For communication service providers voice have been the killer application and it is going to be big source of revenue for years to come. When there were three different voice solutions (IMS-based VoIP, CSFB, VoLGA) there was rightfully serious concerns whether LTE would come with voice anytime soon. So the situation in year 2009 was equally challenging for both communication service providers and vendors. This is why, from time to time, serious rivals come together to agree on technical cooperation that is designed to help smooth the way forward for the common good of the telecoms market. On 4 November 2009 the One Voice initiative was published by AT&T, Orange, Telefonica, TeliaSonera, Verizon, Vodafone, Alcatel-Lucent, Ericsson, Nokia Siemens Networks, Nokia, Samsung and Sony Ericsson. These 12 companies announced that they have concluded that the IMS-based solution, as defined by 3GPP, is the most applicable approach to meet their consumers' expectations for service quality, reliability and availability when moving from existing circuit switched telephony services to IP-based LTE services. The companies in One Voice then set about to create a solid foundation for securing the smooth introduction of standards-based VoLTE. They evaluated the different alternative 'ingredients' specified by 3GPP in order to settle on a minimum set of essential handset and network functionalities and features that communication service providers would need to implement basic, interoperable VoLTE service. These agreed mandatory set of functionalities for the user equipment (UE), the LTE access network, the EPC network and the IMS functionalities are contained in the 'technical profile' published by One Voice, available for use by anyone in the industry. In a sense, the technical profile gives all industry stakeholders a level playing field on which to enhance their VoLTE service as they see fit, but most importantly a level playing field that enables the basic working, and interworking, of VoLTE across the entire industry landscape.

15 February 2010 marks the second important milestone in VoLTE ecosystem development. On that date, the Global System for Mobile Association (GSMA) announced it has adopted the work of the One Voice initiative to drive the global mobile industry towards a standard way of delivering voice and messaging services for LTE (GSMA) and Next Generation Mobile Networks alliance (NGMN) delivered communication service providers' agreement to ensure roaming for VoLTE by recommending to support CSFB in all LTE voice devices and networks (NGMN). The GSMA's VoLTE initiative was supported by more than 40 organisations from across the mobile ecosystem, including many of the world's leading mobile communication service providers, handset manufacturers and equipment vendors, all of whom support the principle of a single, IMS-based voice solution for next-generation mobile broadband networks. This announcement was also supported by 3GPP, NGMN and the International Multimedia Teleconferencing Consortium (IMTC). Following the announcement, work progressed very quickly; and already in March 2010 the GSMA permanent reference document (IR.92) on IMS profile for voice and short message service (SMS) was published containing an improved version of the One Voice profile. In September 2010 GSMA agreed to freeze the content of the permanent reference document (IR.92). A global baseline for commercial VoLTE deployments was finally stabilised.

2

VoLTE Deployment Strategies

Global mobile networks have been driven by voice service since the beginning and voice has been the 'killer' application for many years. It can be said that there would be no mobile networks without voice services. Today, we have a new situation where data traffic is heavily growing in mobile networks and all deployments are going to be driven by data increase. Also from a standards point of view a data-driven approach has been selected where future mobile networks are based on the internet protocol (IP) technology and mobile services are built on top of the 'data layer'.

The third generation partnership project (3GPP) long term evolution (LTE) is the main technology for further mobile networks and it can be said already now that it is the fastest developing mobile system technology ever. One reason is that LTE first deployments are for data with dongles and data cards including short message service (SMS) support, but services like voice will follow later. From a LTE deployment perspective the LTE voice service can be regarded as just one data application, but with specific requirements for real-time traffic, quality of service and interoperation with existing voice infrastructure [circuit switched (CS) core networks]. It is also self-evident that voice remains as a mandatory service in LTE.

The LTE/evolved packet core (EPC) standards have been designed as mobile technology based purely on the IP. The LTE/EPC networks will be driven by mobile broadband services in which the voice and SMS traffic amounts are a smaller portion of the total traffic amount but still retain their importance for both subscribers and operator business. From a technology point of view the LTE/EPC network is designed to carry all applications and it has excellent support for voice service with low latency and high capacity, which means savings because in the end the operator does not have to maintain parallel networks for voice and data.

The voice service over LTE is done with an IP multimedia subsystem (IMS) as specified by 3GPP. LTE radio access does not support direct connectivity to CS core and services, but radio is connected to EPC that provides IP connectivity for the end user services and interworking towards existing CS networks.

2.1 Common Networks Everywhere

It is fair to assume a long migration from CS voice to a full blown IMS-based voice over internet protocol (VoIP) and therefore 3GPP has standardised several building blocks

Voice over LTE: VoLTE, First Edition. Miikka Poikselkä, Harri Holma, Jukka Hongisto, Juha Kallio and Antti Toskala.

to enable different roadmaps towards an all-IP network. With these building blocks a communication service provider can have the right voice introduction based on its own voice strategy, that is more CS reuse or more IMS-driven.

The LTE is a common technology for 3GPP and 3GPP2 operators and from a voice deployment strategy point of view it is good to understand the differences between existing global system for mobile communication (GSM), wideband code division multiple access (WCDMA) and code division multiple access (CDMA) deployments. Both GSM and WCDMA technologies provide simultaneous voice and data service, even though GSM dual transfer mode (DTM) is not as widely used as WCDMA multi radio access bearer (RAB). The CDMA technology is widely limited to evolution data only (EV-DO) where simultaneous voice and data is not possible.

2.2 GSM/WCDMA View

The expected deployment strategy for existing GSM/WCDMA communication service providers is to start with circuit switched fallback (CSFB) and gradually move towards IMS-based VoIP deployments. Notice that either of the interim all-IP steps can be omitted in this evolution. When LTE radio network is introduced the main steps to provide voice service are as follows:

1. CSFB as a first voice solution for LTE subscribers.
2. IMS-based VoIP with single radio voice call continuity (SR-VCC) prior countrywide LTE coverage exists. SR-VCC enables handover from IMS-based VoIP to CS speech when user equipment (UE) is running out of VoIP coverage, and reverse SR-VCC enables handover from CS speech to IMS-based VoIP.
3. IMS-based VoIP where all the calls are made over packet switched networks (all-IP).

In addition to the steps listed above there can also be mixed environments where some voice services use CSFB but others use IMS-based VoIP. For example IMS-based VoIP for a communication service provider's own subscribers and CSFB for inbound roamers.[1]

2.3 CDMA View

The starting point for an existing CDMA operator is that the voice service is offered over a CS network. When introducing LTE radio network the potential steps to provide voice service are as follows:

1. Simultaneous voice and long term evolution (SV-LTE) or simultaneous voice and data only (SV-DO) where dual radio terminal offers simultaneous LTE/evolved high rate packet data (eHRPD) data and CDMA 1×voice. This allows simultaneous voice and data service use for CDMA users.
2. CSFB is also an option for some CDMA operators, but is not necessary needed if mobiles offer SV-LTE.
3. Eventual target IMS controlled voice over all-IP network where all the calls are made over packet switched networks, either on the LTE only or on the LTE and eHRPD.

[1] The CSFB will be used as the initial voice roaming solution because IMS based VoIP roaming is still being finalised in GSMA and 3GPP at the time of writing.

Single radio voice call continuity is also standardised from LTE to single carrier radio transmission technology (1×RTT) but there exists the same limitation as on CSFB, that is a simultaneous data and voice service cannot be offered after UE is moved to 1×RTT during a voice call. In general, it can be said that both CSFB and SR-VCC are possible but they require such investment to legacy CDMA technology that the operators' interest seems to be rather full LTE coverage as soon as possible with IMS based VoIP. The intermediate phase will passed mostly with SV-LTE terminals, and CSFB will be used only by some operators.

3

VoLTE System Architecture

3.1 Overview

This chapter introduces the reader to the voice over long term evolution (VoLTE) system architecture. This includes various network entities, how they interact with each other via different reference points. In this book we split architecture into three domains: access, evolved packet core (EPC) and control, as shown in Figure 3.1.

Figure 3.1 reveals that access contains only one network element, the so-called evolved NodeB (eNodeB or eNB). In previous third generation partnership project (3GPP) radio architectures two networks elements were defined: (i) NodeB and radio network controller (RNC) in the UMTS terrestrial radio access network (UTRAN) and (ii) base transceiver station (BTS) and base station controller (BSC) in the GSM/EDGE radio access network (GERAN). The main reasons for this simplified architecture were reducing complexity, latency and costs while increasing data throughput which we will further explain in Section 3.2.

The EPC consists basically of three functional elements. The first one is the mobility management entity (MME) that resides in the control plane of EPC. The MME can be seen as an evolution of the serving GPRS support node (SGSN) control plane function in the general packet radio system (GPRS). The serving gateway (S-GW) correlates with the SGSN user plane function in GPRS. All user plane packets in uplink and downlink are traversing the S-GW and the S-GW also acts as a local mobility anchor that is able to forward packets during handover. The PDN gateway (P-GW) finally is the global Internet protocol (IP) mobility anchor point comparable to the gateway GPRS support node (GGSN) in GPRS. It allocates IP addresses to user equipments (UEs) and provides the interface towards packet data networks (PDNs) like the IP multimedia subsystem (IMS) and Internet. The P-GW contains also the policy and charging enforcement function (PCEF) for the detection of service data flows, policy enforcement (e.g. discarding of packets) and flow-based charging. Section 3.3 details further EPC architecture.

The control domain contains three building blocks: (i) IMS, (ii) home subscriber server (HSS) and (iii) policy and charging rules function (PCRF). IMS further contains number of functional elements which are required to control voice sessions. The HSS is the main data storage for all subscriber and service-related data of the IMS, EPC and access. The PCRF acts as a bridge between IMS and EPC domains, for example setting up required EPC bearer for a new voice session. You will find a description of all control domain entities in Section 3.4.

Voice over LTE: VoLTE, First Edition. Miikka Poikselkä, Harri Holma, Jukka Hongisto, Juha Kallio and Antti Toskala.

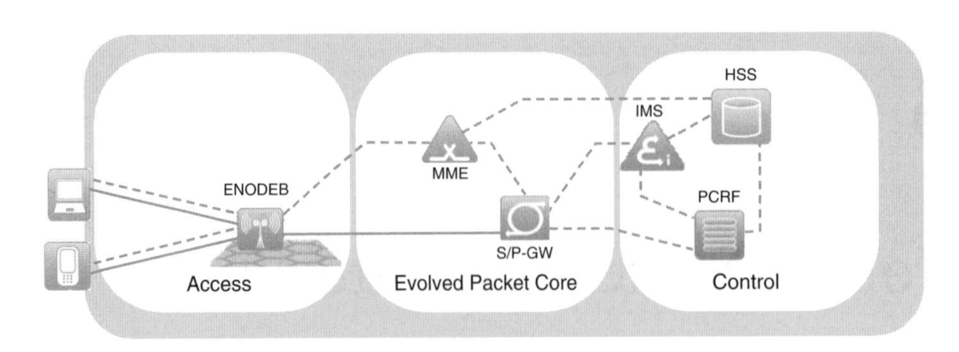

Figure 3.1 VoLTE system architecture.

3.2 LTE Radio

In this section we introduce long term evolution (LTE). An overview is given on the developed LTE; and the key design principles are presented. Also an overview of the key radio functionality and the related importance to the provided VoLTE is introduced.

3.2.1 LTE Radio Background

The LTE radio development in 3GPP was formally started in 2005, a bit before the demand for the data capacity started to increase rapidly. Inline with the name LTE, the LTE was intended to address the longer term development of the radio technology in order to have next generation radio access as part of the 3GPP radio family. During the definition of LTE radio specifications, the demand of wireless packet data services grew rapidly with the success of high speed packet access (HSPA) data solutions. This development was giving clearer motivation for a new efficient packet radio system optimised for packet data, which had been the design criteria from the start. The content for Release 8 was finalised in the end of 2008, as shown in Figure 3.2, followed by the protocol freeze for the protocol between UE and the network in March 2009. 3GPP then continued development with further Releases, with Release 9 being completed one year later, being a rather small

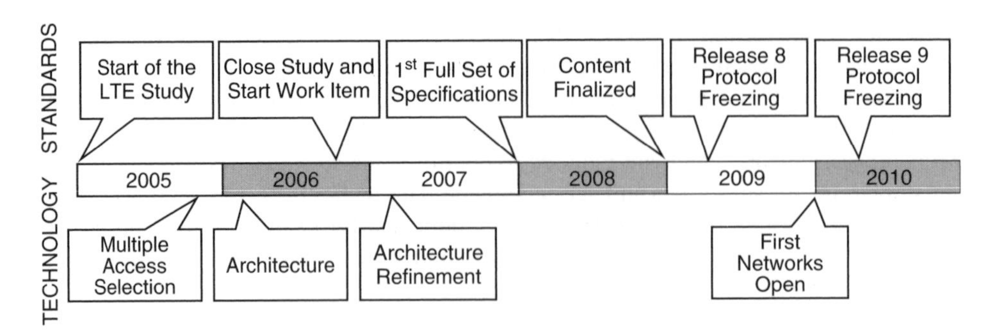

Figure 3.2 LTE development time line.

Release but with important small additions, especially for the voice deployment, such as positioning support, as covered in more detail in Section 4.2.6. Release 10, finalised in June 2011, is rather focused on increasing the LTE capabilities for high speed data operation with LTE advanced, and as such it does not contain new essential features for the support of voice in LTE. For more details on LTE advanced see (Holma, H. and Toskala, A. "LTE for UMTS", 2nd edition, Wiley, 2011).

3.2.2 LTE Radio Architecture

The LTE architecture is based on the principle of a single box radio architecture. The classical controller-based architecture has evolved to the 'all in BTS' radio architecture, which was also introduced as an alternative architecture in HSPA. The key driver was to enable easy scalability of the network as traffic increases, coupled with the separate handling of control and user data in the core network side, as shown in Section 3.1. The use of flat architecture also enables short latency and fast signalling exchange since all radio level signalling is only between the UE and eNodeB. As shown in the LTE architecture (Figure 3.3), in order to facilitate radio resource management and especially mobility, there is an X2 interface between the eNodeBs. This interface carries the necessary signalling to enable to exchange information on the radio resource usage between the base station, to provide handover command when user is moving or to handle the data forwarding

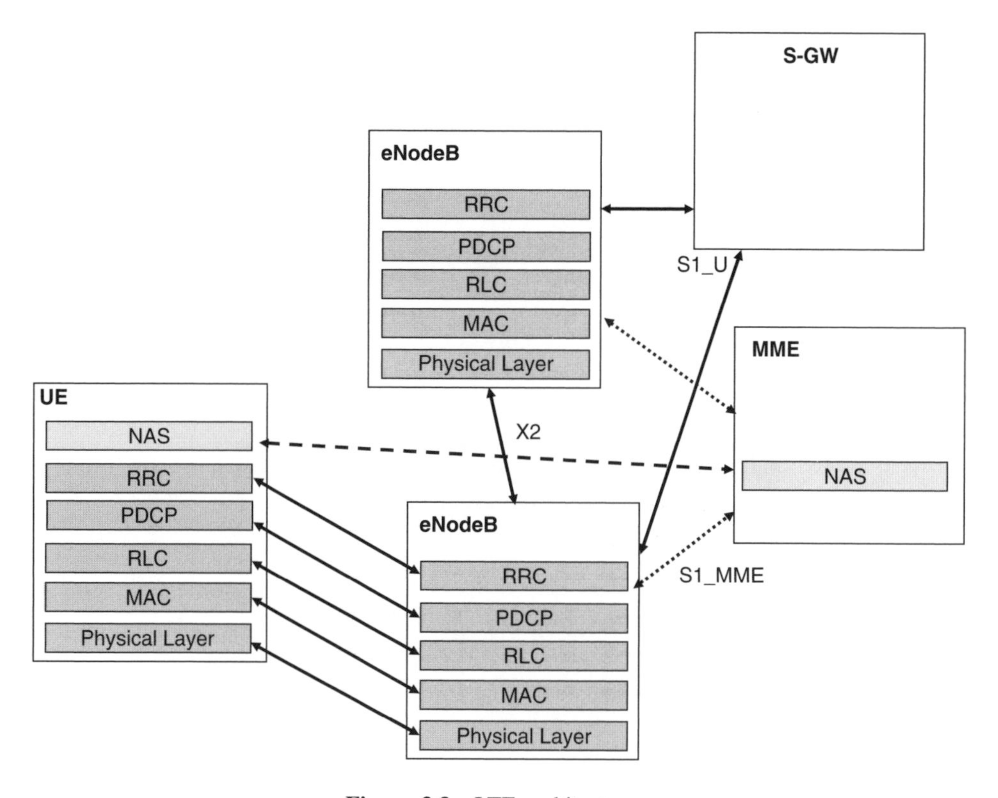

Figure 3.3 LTE architecture.

when user is moving to another eNodeB. Some of the functions, like handover command or data forwarding are also possible via core network. Towards the core network there is connection via S1_U for the user data (towards the S-GW in the core network) and S1_MME for the core network related control signalling (towards the MME). This allows scaling the core network processing capability separately as a function of data volume and as the function of the amount of users in the network.

The LTE protocol architecture shows that all radio related protocols terminate in the eNodeB, that is responsible for all radio resource related signalling and decisions. The non-access stratum (NAS) signalling, terminating in the MME, refers to non-radio signalling, such as authentication at the evolved packet system (EPS) attachment. The protocol stack further from the eNodeB is shown in Figure 3.4. The user data is carried on top of the user datagram protocol (UDP) while the control plane data uses the stream control transmission protocol (SCTP) for better reliability because SCTP also ensures in-sequence delivery of the messages. The GPRS tunnelling protocol (GTP-U) running on top of UDP carries the user data to the serving gateway (S-GW). The S1 application Protocol (S1AP) is the signalling protocol between MME and eNodeB.

The further connections from the S-GW and MME are shown in Figure 3.7. The connection towards the circuit switched (CS) core network for provision of the user data from the CS core does not exists in LTE, as LTE has been intended for the packet based of communication only, including voice as well, thus relaying on the use of the IMS

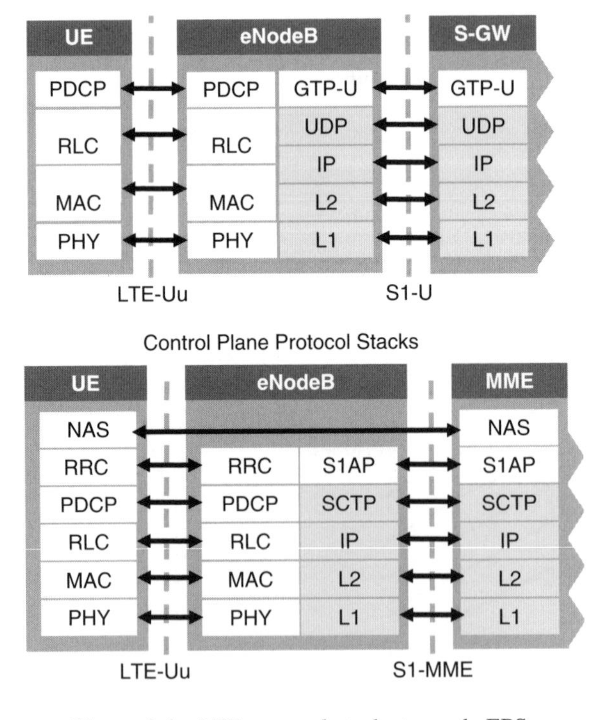

Figure 3.4 LTE protocol stacks towards EPS.

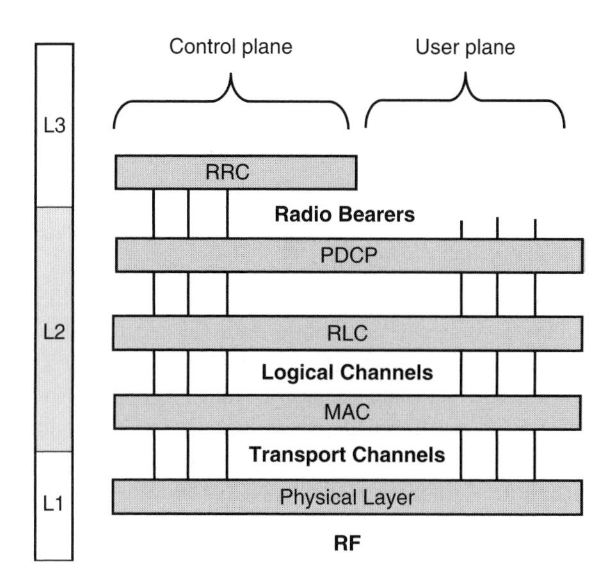

Figure 3.5 LTE radio protocol layer.

for voice service delivery. There is a possibility to have a signalling connection between the MME and Mobile Switching Centre (MSC) server as discussed in Section 4.2.3 in connection with the circuit switched fallback (CSFB) procedure.

The LTE radio protocol stack in Figure 3.5 has the following key functionality:

- The radio resource control (RRC) handles all necessary control signalling for the radio operation, including measurement control and handover commands.
- The packet data convergence protocol cover the ciphering and also IP header compression. Use of header compression is not mandatory but quite essential in order to achieve good voice capacity as otherwise the IP header overhead takes quite a large part of the capacity in the case of large amounts of voice user.
- The radio link control (RLC) does the retransmission control, though this not normally used for voice. It also does the segmentation.
- The medium access control (MAC) functions include the mapping and multiplexing of logical channels to transport channels as well as taking care of the physical layer retransmissions (which can be also applied to VoLTE due to the low delay operation). The scheduling functionality with priority handling is also seen as MAC layer functionality, thus ensuring that VoLTE users do not have too much delay but can have high priority in the scheduler operation in case of congestion situation.
- The physical layer covers the channel encoding and decoding, error detection and actual physical layer packet combining modulation/demodulation and many other procedures for power control, and multi-antenna operation. The physical layer is based on the orthogonal frequency division multiple access (OFDMA) for downlink and single carrier frequency division multiple access (SC-FDMA) in the uplink to ensure both simple terminal receiver and power-efficient terminal power amplifier operation while reaching high capacity, as analysed in detail in (Holma, H. and Toskala, A. "LTE for UMTS", 2nd edition, Wiley, 2011).

3.3 Evolved Packet Core

In this section we introduce the EPC. We provide an overview of packet core development history and we explain EPC entities and key procedures: mobility and session management.

3.3.1 What is the Evolved Packet Core?

The EPC is continuous from the GPRS, which is used for mobile data services. The GPRS was first a packet switched (PS) mobile data service introduced in the European Telecommunications Standards Institute (ETSI)/3GPP release 97. The main difference between PS data service and CS data service is that PS data service does not have a constant resource reservation and the usage is volume-based rather than time-based.

In GPRS there is a dedicated core for PS data service; the PS core that consists of SGSN and GGSN providing 'always on' IP-based connectivity to data services.

The first GPRS release was improved with higher data rates in ETSI/3GPP Release 98, and the first universal mobile telephone system (UMTS) release was introduced in ETSI/3GPP Release 99. In Release 99 UMTS network there came new access technology and a new interface from the PS core to the radio access network (RAN). The next 3GPP releases introduced higher bit rates and network optimisations like Direct Tunnel in 3GPP Release 7.

In 3GPP Release 8 whole system was re-designed under the system architecture evolution (SAE) work item. As a result of this work the EPC was introduced. While the PS core in GPRS is used for global system for mobile communication (GSM) and UMTS access technologies the EPC can be used for GSM, UMTS, LTE and non-3GPP access technologies to provide IP connectivity and service continuity in mobile networks.

One difference between the GPRS core and EPC is the strict and in-built separation of control and user plane in EPC. This is possible in GPRS as well by using the direct tunnel feature, but with direct tunnel the SGSN user plane is used, for example, in an international roaming scenario.

For LTE the EPC contains two functional elements in the user plane, the S-GW and P-GW, and one element in the control plane, the MME. As a consequence, a minimum of two functional elements (eNodeB, combined S-GW/P-GW) are in the EPS user plane path, which means a higher degree of simplicity and less latency.

Figure 3.6 shows 3GPP architecture evolution for a mobile PS network and how a number of user plane elements are reduced.

3.3.2 EPC Entities and Functionalities

3GPP Release 8 defines EPC entities and functionalities. Figure 3.7 shows standard Release 8 architecture with interfaces between logical entities.

The MME has a similar role to SGSN in 2G/3G GPRS networks. 3GPP Release 8 mobile gateway functionality is divided into S-GW and P-GW functionalities and they are connected via an open S5 interface.

A LTE subscriber is always connected to services via the P-GW even though user would be using 2G/3G radio access. The S-GW terminates the user plane interface towards the

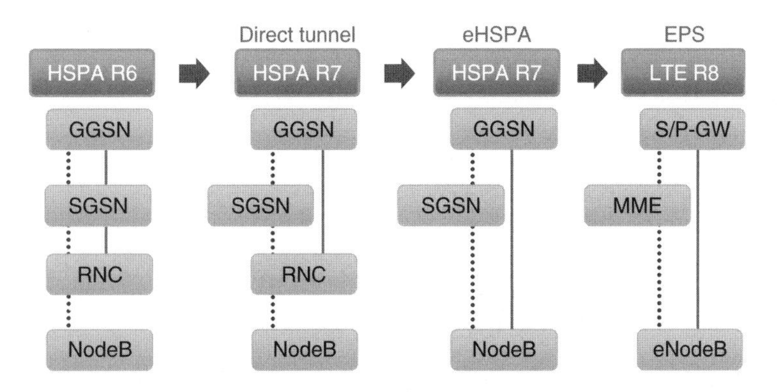

Figure 3.6 Mobile packet switched network evolution. Reproduced with permission from John Wiley & Sons, Ltd.

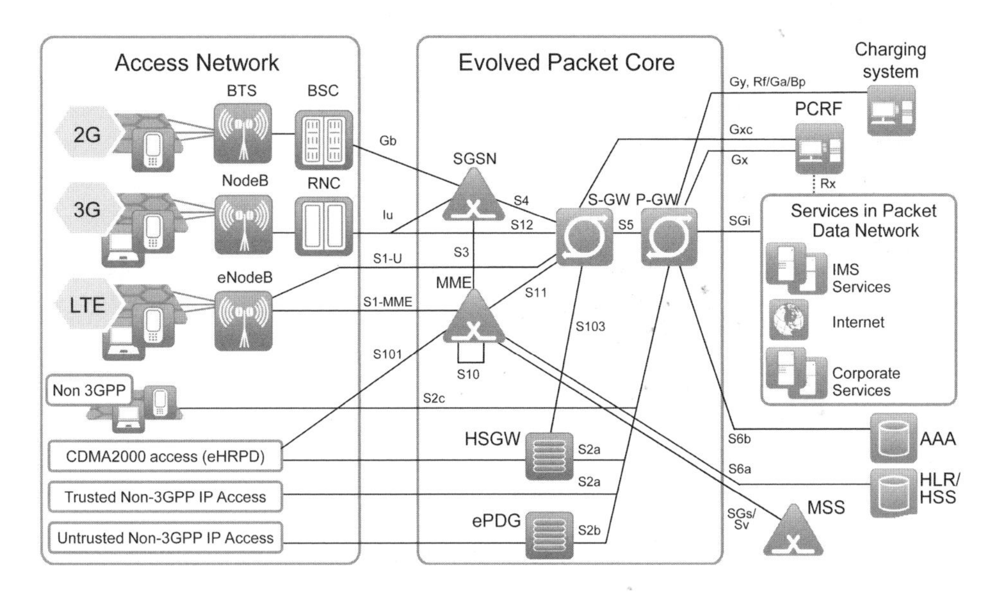

Figure 3.7 3GPP Release 8 reference architecture.

evolved UMTS terrestrial radio access network (E-UTRAN). Each UE is assigned to a single S-GW at a given point of time. The S-GW acts as a user plane gateway for LTE radio network in inter-eNodeB handovers and for inter-3GPP mobility (terminating S4 and relaying the traffic between 2G/3G system and PDN-GW). The P-GW acts as IP point of attachment of UE and terminates the SGi interface towards the service networks.

MME is, as its name says, a management entity and a main control element for mobility ina LTE network. The MME main tasks are:

- NAS signalling and security;
- Authentication and authorisation;

- Session and mobility management, including intra LTE and intra 3GPP mobility;
- UE paging control and execution;
- Tracking area list management;
- Inter (CN) node signalling;
- MME/SGSN selection for handovers;
- S-GW and P-GW selection and S-GW relocation;
- Roaming (S6a towards home HSS);
- Lawful interception for signalling traffic;
- CSFB for mobility from LTE to CS;
- (SRVCC) for mobility from LTE to CS.

S-GW is a user plane 'proxy' for 3GPP mobility over 2G, 3G and LTE accesses. The S-GW main tasks are:

- User plane anchor for inter-eNodeB mobility and handovers;
- User plane anchor for mobility between 2G/3G and LTE;
- Assist the eNodeB during inter-eNodeB handover (data forwarding);
- Downlink packet buffering and initiation of triggering for paging idle mode UEs;
- Packet routing and forwarding;
- Transport level downlink and uplink packet marking;
- Lawful interception for roaming;
- Charging for data roaming.

P-GW is a user plane anchor between 3GPP and non-3GPP accesses. The P-GW main tasks are:

- Gateway towards PDNs (i.e. Internet/intranets/operator services);
- Packet routing and forwarding;
- Bearer management;
- UE IP address allocation;
- Policy and Charging Enforcement Function (PCEF);
- Accounting and offline/online charging support for data;
- Packet filtering (optionally deep packet inspection);
- Lawful interception;
- User plane anchor for mobility between 3GPP and non-3GPP access systems.

Besides MME, S-GW and P-GW, a couple of other logical functions are part of the complete EPS:

- The SGSN shown in Figure 3.7 (also called S4-SGSN) is enhanced to support EPS bearer model and 3GPP Release 8 procedures towards MME and S-GW. The S4-SGSN behaves like the MME towards the EPC but provides a Iu interface to UTRAN and a Gb interface to GERAN.
- The HSS contains all subscription relevant data of the users like international mobile subscriber identifier (IMSI), mobile station integrated services digital network (MSISDN), subscribed access point names (APN) and subscribed quality of service (QoS) and so on. Please see the overall description in Section 3.4.5.

- The PCRF provides charging and policy rules to P-GW and translates application level session data coming from the application server [AS; for example, the content of session description protocol (SDP) in session initiation protocol (SIP) signalling] into 3GPP specific parameters. Please see the overall description in Section 3.4.6.
- The authentication, authorisation and accounting (AAA) server is used to authenticate and authorise users who are accessing from non-3GPP access systems or, optionally, by the P-GW to authorise the provided APN and to allocate an IP address to the UE.
- The evolved packet data gateway (ePDG) terminates a secure tunnel from the UE in the EPC when the UE is camping on an untrusted non-3GPP access system.
- The online charging system (OCS) performs real-time credit control. Its functionality includes, for example transaction handling, rating, online correlation and management of subscriber accounts and balances.
- The offline charging system (OFCS) collects charging relevant data from network elements and passes them to the operator's billing domain to generate subscriber billing records.
- The HRPD serving gateway (HSGW) is a 3GPP2 specified access gateway, which provides connectivity from evolved high rate packet data (eHRPD) when LTE capable UE is camping in code division multiple access (CDMA) data access.

3.3.3 EPS Mobility Management

Here we attribute EPS mobility management (EMM) into four different parts:

1. UE registration via attach and de-registration via detach (see more about the initial attach procedure in Section 5.3);
2. Maintaining signalling connection for active UE during mobility;
3. Tracking UE when the signalling connection is not established for registered UE;
4. Re-establishing signalling connection when the UE becomes active.

The EMM states (see also Figure 3.8) describe the mobility management and the EPS connection management (ECM) states describe the signalling connectivity between the UE and the EPC (see also Figure 3.9).

When UE is EMM-DEREGISTERED the UE is not reachable by a MME but UE context can still be stored in the UE and MME. When UE is EMM-REGISTERED the UE location is known an accuracy of the tracking area (list) and UE has at least one active PDN connection.

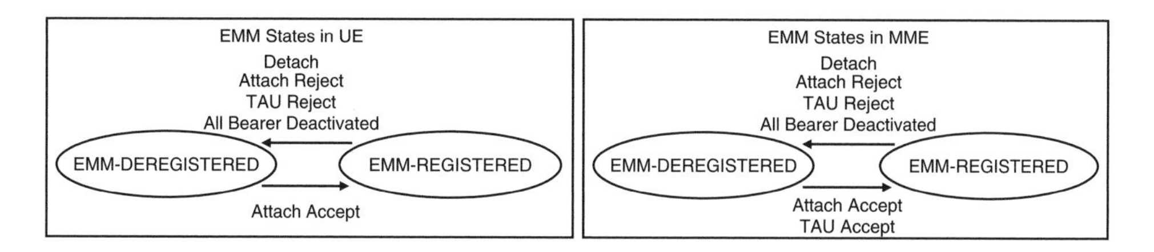

Figure 3.8 EPS mobility management states.

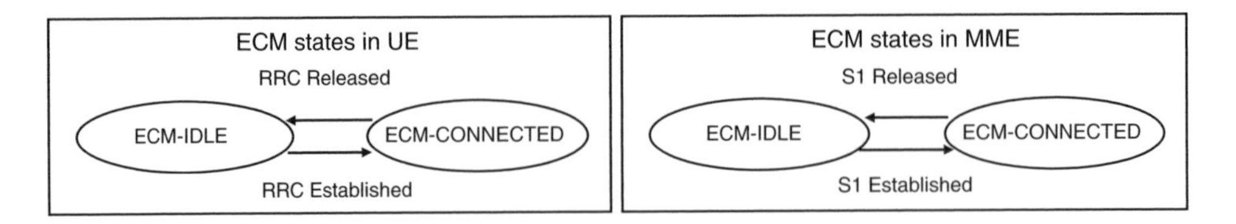

Figure 3.9 EPS connection management states.

UE is in ECM-IDLE state when no NAS signalling connection between UE and network exists. In the EMM-REGISTERED and ECM-IDLE state, the UE performs the following: (i) tracking area update, (ii) periodic tracking area update, (iii) service request and (iv) answer the paging from MME with service request. UE and MME enter the ECM-CONNECTED state when signalling connection is established. In the ECM-CONNECTED state the UE location is known in the MME with the accuracy of a serving eNodeB. The S1 release procedure changes the state at both UE and MME from ECM-CONNECTED to ECM-IDLE.

3.3.3.1 Tracking Area Update

Once the UE is successfully registered, it needs to keep the network informed about its current location in order to be reachable for downlink signalling and user data even when the UE is in idle mode, that is when the signalling connection to the UE has been released.

For this purpose E-UTRAN cells are combined to so-called tracking areas. A UE camping on an E-UTRAN cell in idle mode is listening to the system information broadcast in this cell, which includes the tracking area identity of the cell. When the UE is changing to a new cell and the received tracking area identity indicates that the new cell belongs to a tracking area to which the UE is currently not registered, the UE initiates a tracking area updating procedure.

The tracking area or list of tracking areas allocated by the MME during the tracking area updating procedure or attach procedure can be used by the MME to page the UE in a certain area, when the signalling connection to the UE has been released and the network needs to send downlink signalling or user data.

3.3.3.2 Intra LTE Handovers

When the UE is moving in active state, for example during a voice call, it can be handed over from the currently serving eNB to a new eNB. The handover procedure can be either a X2-based or a S1-based handover, named after the interface used for the exchange of signalling messages during the handover preparation.

The X2-based handover is more optimised and useful intra LTE handover procedure for the real-time traffic. The handover preparation and execution is done between the radio access elements, and the EPC role in the X2 handover is to modify the bearer(s) path after handover execution. The S1-based handover procedure is needed when the X2-based handover cannot be used or MME must be changed. X2-based handover is depicted in Figure 3.10.

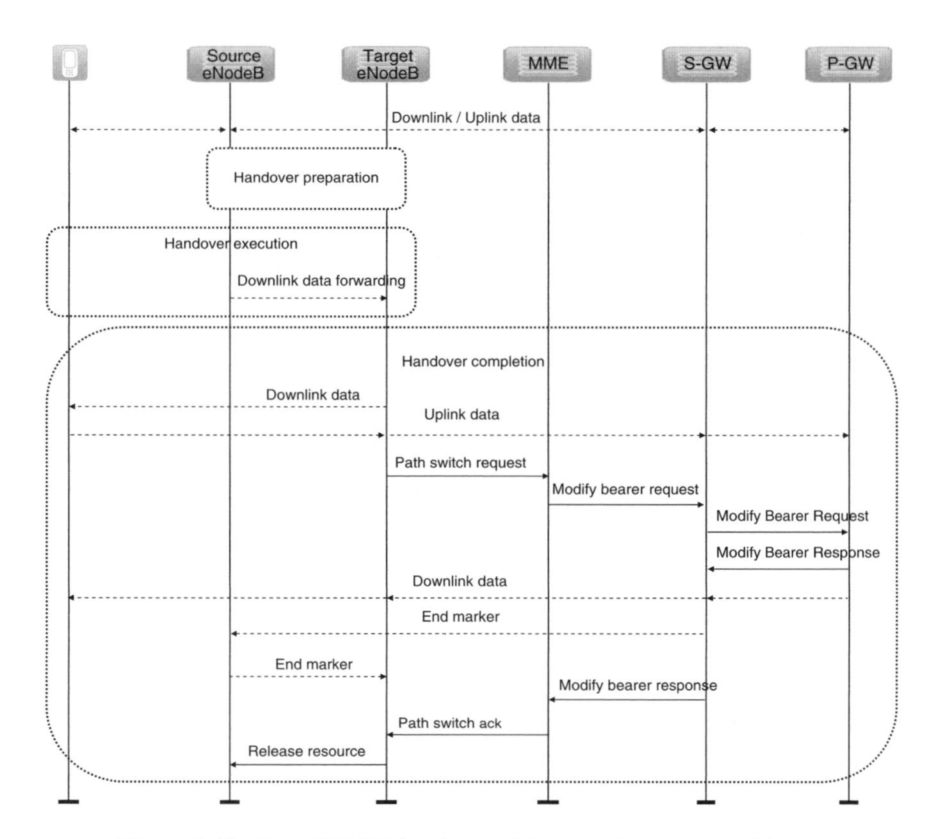

Figure 3.10 Intra LTE X2 handover without serving gateway change.

Handover preparation and execution is done between source eNodeB and target eNodeB. During the handover preparation there is ongoing data transfer for the active voice call. In handover execution the source eNodeB commands UE to the target side and downlink packets are forwarded by the source eNodeB to the target eNodeB.

After UE is moved to the target eNodeB the user plane path is switched to the target eNodeB and bearer information is updated in EPC. By sending a *path switch request* the target eNodeB indicates to the MME that the S1 interface needs to be switched from the source eNodeB to the target eNodeB. The MME updates the S-GW with the new eNodeB address information for the downlink of the user plane and confirms the relocation of the S1 interface towards the target eNodeB with a *path switch ack (acknowledgement).* The MME may also allocate a different S-GW and provide the target eNB with the necessary address information for the uplink of the user plane in the path switch ack message. The S-GW may provide a new radio access location to P-GW, or if S-GW is changed then a new S-GW address needs to be signalled to P-GW.

After completion of the handover the UE receives the system information broadcast in the target cell. If the target cell of the handover belongs to a tracking area in which the UE is currently not registered, the UE initiates a tracking area updating procedure. See also Section 4.2.2 for intra-frequency mobility.

3.3.4 EPS Session Management and QoS

IP connectivity between a UE and a public land mobile network (PLMN) external PDN is referred to as PDN connectivity service. Each PDN connection consists of one or more bearers that carry data over the EPS, as shown in Figure 3.11. An EPS bearer is established when the UE connects to a PDN and that remains established throughout the lifetime of the PDN connection to provide the UE with always-on IP connectivity to that PDN. That EPS bearer is referred to as the default bearer (see also Table 3.1 for the key characteristics of a default bearer). Any additional EPS bearer that is established for the same PDN connection is referred to as a dedicated bearer (see also Table 3.1). For IMS-based services like VoLTE, there must be always an PDN connection established for the IMS APN before IMS registration.

The UE routes uplink packets to the different EPS bearers based on uplink packet filters in the traffic flow templates (TFTs) assigned by the P-GW. The P-GW routes downlink packets to the different EPS bearers based on the downlink packet filters in the TFTs. For each bearer there is a GTP-U tunnel between eNodeB and S-GW, and between S-GW and P-GW. Bearer service signalling is done with GPRS tunnelling protocol – control plane version 2 (GTP-Cv2) inside the core and with S1-AP between MME and eNodeB.

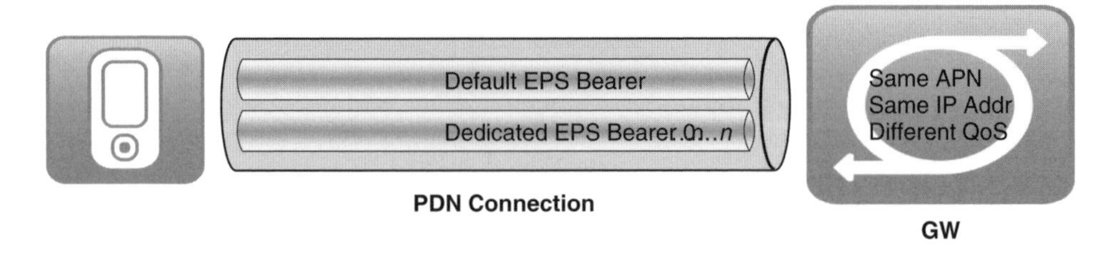

Figure 3.11 Packet data network connection including EPS bearers.

Table 3.1 Default bearer versus dedicated bearers

Default bearers	Dedicated bearers
One default bearer is created when UE attaches to LTE and at least one default bearer remains active as long as UE is attached to LTE → always on	Dedicated bearer is created for QoS differentiation purposes and triggered by the network or UE
Additional default bearers (and thus PDN connections) may be created when simultaneous access to services available via multiple APNs is needed, for example *Internet* APN for internet services and *IMS* APN for IMS services	Dedicated bearers are controlled and established by the network based on the operator policy
Default bearers are always non-GBR bearers	Dedicated bearers are either non-GBR bearers or GBR bearers

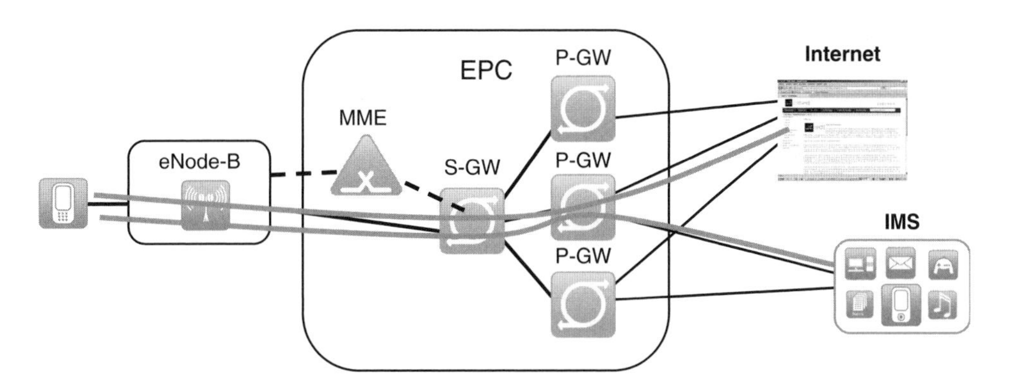

Figure 3.12 PDN connections via IMS and internet APNs.

An additional default bearer(s) with a new IP address is created when services are offered via multiple APN. For example UE gets PDN connection to *IMS* APN during the initial attachment, but when the user starts web browsing the UE initiates another PDN connection to *Internet* APN. Additional default bearers are always initiated by the UE request and removed by the UE or network.

Figure 3.12 shows an example where UE has two simultaneous PDN connections, one for IMS and another for the Internet.

Note that different APNs may be supported by different P-GWs. For example the *IMS* APN might be supported only in the limited number of P-GWs, whereas the Internet APN might be supported in all P-GWs.

3.3.4.1 IP Address Allocation

UE gets at least one IP address [i.e. IP version 4 (IPv4) address, IPv6 prefix or both] during the PDN connection establishment. The same address is also used for the default bearer and dedicated bearer(s) within the same PDN connection. The IP address can be allocated using one of the following ways:

- The home public land mobile network (HPLMN) allocates the IP address to the UE when the default bearer is activated. This may be a dynamic or a static HPLMN address.
- The visited public land mobile network (VPLMN) allocates the IP address to the UE when the default bearer is activated. This is a dynamic VPLMN address.
- The PDN operator or administrator allocates an IP address to the UE when the default bearer is activated.

3.3.4.2 Quality of Service

The EPS provides bearer level quality of service (QoS), which means that all traffic mapped to the same EPS bearer receives the same bearer level packet forwarding treatment. Providing different bearer level packet forwarding treatment requires separate EPS bearers. The EPS bearer uniquely identifies traffic flows that receive a common QoS treatment between a UE and a P-GW (GTP-based S5/S8). The EPS bearer TFT is the set of all packet filters associated with that EPS bearer.

A dedicated bearer [non-GBR or GBR (guaranteed bit rate)] is created for QoS differentiation purposes. The IP address(es) and APN allocated for the default bearer are used for the dedicated bearers within the same PDN connection. Creation of a dedicated bearer is initiated by network, but may be triggered by UE request for resources (NW decides whether existing bearer can be used or new one is needed). A more detailed description for dedicated bearer establishment is described in Section 4.3.3.

The distinction between default and dedicated bearers is transparent to eNodeB. An EPS bearer is referred to as a GBR bearer if the dedicated network resources related to a GBR value associated with the EPS bearer are permanently allocated at bearer establishment/modification. Otherwise, an EPS bearer is referred to as a Non-GBR bearer.

For each EPS bearer (default and dedicated), QoS support is based on QoS parameters:

- **QoS class identifier (QCI)**: Scalar that is used as a reference to access node-specific parameters that control bearer level packet forwarding treatment.
- **Allocation and retention priority (ARP)**: Contains information about the priority level (scalar), the pre-emption capability (flag) and the pre-emption vulnerability (flag). The primary purpose of the ARP is to decide whether a bearer establishment/modification request can be accepted or needs to be rejected due to resource limitations.
- **GBR**: Denotes the bit rate that can be expected to be provided by a GBR bearer.
- **Maximum bit rate (MBR)**: Limits the bit rate that be expected to be provided by a GBR bearer.

The following QoS parameters are applied to aggregated set of EPS bearers:

- **APN–aggregate maximum bit rate (AMBR)**: Limits the aggregate bit rate that can be expected to be provided across all non-GBR bearers and across all PDN connections of the APN.
- **UE-AMBR**: Limits the aggregate bit rate that can be expected to be provided across all non-GBR bearers of a UE.

Table 3.2 shows standardised QoS characteristics where VoLTE related QCIs are highlighted.

3.4 Control

In this section we introduce the IP multimedia subsystem (IMS), policy and charging rule function (PCRF) and home subscriber server (HSS). We will explain how these functions are connected and the key protocols used. In addition we will provide an overview of IMS development history and explain IMS design principles.

3.4.1 What is an IP Multimedia Subsystem?

In order to communicate, IP-based applications must have a mechanism to reach the correspondent. The telephone network currently provides this critical task of establishing a connection. By dialling the peer, the network can establish an ad hoc connection

Table 3.2 EPS quality of the service characteristics

QCI	Resource type	Priority	Packet delay budget (ms)	Packet error loss rate	Example services
1	GBR	2	100	10^{-2}	Conversational voice
2		4	150	10^{-3}	Conversational video (live streaming)
3		3	50	10^{-3}	Real time gaming
4		5	300	10^{-6}	Non-conversational video (buffered streaming)
5	Non-GBR	1	100	10^{-6}	IMS signalling
6		6	300	10^{-6}	Video (buffered streaming) Transmission control protocol (TCP)-based (e.g. www, e-mail, chat, ftp, p2p file sharing, progressive video, etc.)
7		7	100	10^{-3}	Voice Video (live streaming) interactive gaming
8		8	300	10^{-6}	Video (buffered streaming) TCP-based, for example www, e-mail, chat, ftp, p2p file
9		9			Sharing, progressive video and so on

between any two terminals over the IP network. This critical IP connectivity capability is offered only in isolated and single-service provider environments in the Internet; closed systems compete on user base, where user lock-in is key and interworking between service providers is an unwelcome feature. Therefore, we need a global system – the IMS. It allows applications in IP-enabled devices to establish peer to peer and peer to content connections easily and securely. Our definition for the IMS is:

> IMS is a global, access-independent and standard-based IP connectivity and service control architecture that enables various types of multimedia services to end users using common Internet-based protocols.

3.4.2 IMS Development History

The last GSM-only standard was produced in 1998, and in the same year the 3GPP was founded by standardisation bodies from Europe, Japan, South Korea, the USA and China to specify a 3G mobile system comprising wideband code division multiple access (WCDMA) and time division/code division multiple access (TD-CDMA radio access and an evolved GSM core network (www.3gpp.org/About/3gppagre.pdf). It took barely a year to produce the first release – Release 1999. The functionality of the release was frozen in December 1999 although some base specifications were frozen afterward – in March 2001.

After Release 1999, 3GPP started to specify Release 2000, including the so-called all-IP that was later renamed as the IMS. This all-IP system was expected to provide full replacement of mobile CS communication system and to be the service engine of wide range of multimedia applications. During 2000 it was realised that the development of IMS could not be completed during the year. Therefore, Release 2000 was split into Release 4 and 5. It was decided that Release 4 would be completed without the IMS.

Release 5 finally introduced the IMS as part of 3GPP specifications. The content of Release 5 was heavily discussed and, finally, the functional content of 3GPP Release 5 was frozen in March 2002. The consequence of this decision was that many features were postponed to the next release – Release 6. After freezing the content, the work continued and reached stability at the beginning of 2004. Release 5 defines a finite architecture for SIP-based IP multimedia service machinery. It contains a functionality of logical elements, a description of how elements are connected, selected protocols and procedures. In addition, it is important to realise that optimisation for the mobile communication environment has been also designed in the form of user authentication and authorisation based on mobile identities, definite rules at the user network interface for compressing SIP messages and security and policy control mechanisms that allow radio loss and recovery detection. Moreover, important aspects from the communication service provider point of view are addressed while developing the architecture, such as the charging framework and policy and the service control.

Although, Release 5 basically enabled simplistic mobile voice over Internet protocol (VoIP) support it was not really commercially feasible as the available radio performance was poor and feature wise IMS VoIP was behind voice in the CS domain. That is why a new direction for IMS was set in Release 6 which was completed in September 2005. The Release 6 introduced standardised enhancements for services such as routing and signalling modifications, for example public service identity, sharing a single user identity between multiple devices. Improvements in routing capabilities smoothed the road to complete new standardised services such as presence, messaging, conferencing, push to talk over cellular (PoC). In addition, IMS-CS voice interworking and wireless local area network (WLAN) access to IMS were completed. Moreover, improvements in security, policy, charging control and overall architecture were also completed. At the time of writing it can be said these standardised IMS applications have not taken off commercially in the beginning of 2011. However, Release 6 set the baseline for IMS based service suite which is today known as rich communication suite. In mobile world congress 2011, five big communication service providers shared their plans for launching commercially enhanced rich communication suite.

While 3GPP has finalised its Release 5 and 6, other standardisation development organisations have done parallel developments to define their IMS variants which was a testimony of the growing interest in IMS. The most notable development organisations having their own variants were ETSI telecommunications and Internet converged services and protocols for advanced networking (TISPAN), 3GPP2 and Cablelabs. The downside of this trend was the risk that a potential fragmentation of the IMS market could have a direct impact on time to market, research and development cost. Luckily the industry took decisive steps towards a harmonised IMS, the common IMS. The so-called ‘common

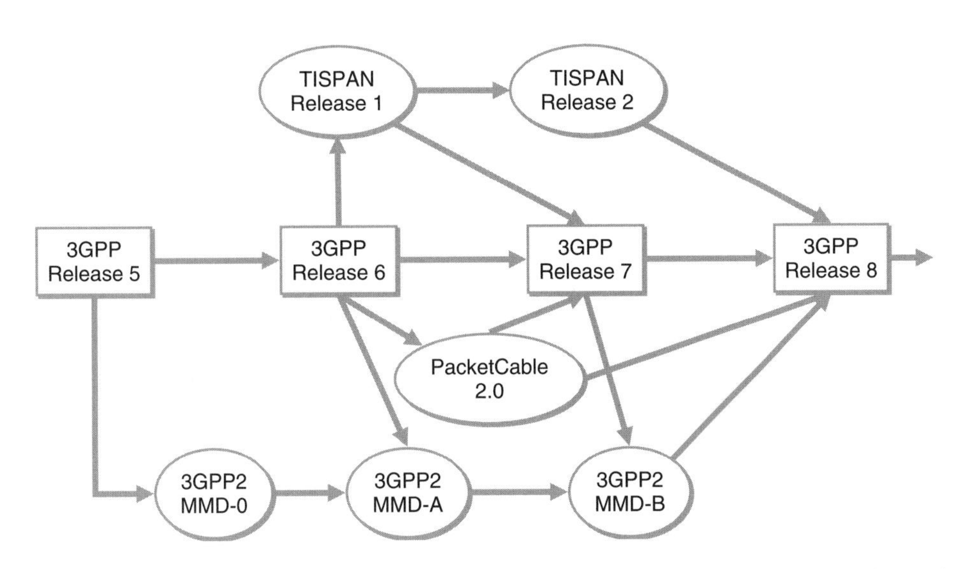

Figure 3.13 Road to standardised common IMS standard. Reproduced with permission from John Wiley & Sons, Ltd.

IMS' agreement was reached between 3GPP and TISPAN, 3GPP2 and Cablelabs to perform IMS related standardisation primarily in 3GPP; 3GPP in turn agreed to take into account requirements and previously developed standards from those other organisations within 3GPP standards. The first round of harmonisation was completed in Release 7 and the effort was effectively completed in Release 8, as depicted in Figure 3.13.

3GPP Release 7 functional content was frozen in March 2007. It introduces two more access technologies [data over cable service interface specification (DOCSIS; the access technology of Cablelabs) and digital subscriber line (xDSL; the access technology of TISPAN)] and features and procedures originating from those and other general improvements. Major new features in Release 7 are: IMS multimedia telephony including supplementary services, short message service (SMS) over any IP access, dual radio voice call continuity, local numbering, combining CS calls and IMS sessions, transit IMS, interconnection border control function (IBCF), globally routable user agent's universal resource identifier (GRUU), IMS emergency sessions, identification of communication services in IMS and a new authentication model for fixed access. Fixed network IMS deployments providing voice started to emerged after completion of Release 7.

Release 8 provides a smooth evolution from the previous releases. The main new building blocks/capabilities are: IMS centralised services [enabling the use of IMS service machinery even though devices are using CS connection (GSM/3G CS radio) towards the network], multimedia session continuity (improving voice call continuity feature to enable continuity of multimedia media streams when IP access is changed), Single-radio voice call continuity [enabling seamless voice handover from LTE (IMS VoIP) to CS], corporate access to IMS [enabling the integration of IP-private branch exchange (PBX) into the IMS network], number portability, carrier selection, IMS inter-communication service provider service interconnection interface, media server architecture enhancements, additional supplementary services [such as communication waiting (CW) and customised

3GPP Release	Expection	Reality
Release 2000	All-IP, CS replacement, 21st century service engine	Release split Rel-4 and no IMS
Release 5	All-IP & service engine of this century	IP connectivity engine
Release 6	Near real-time operator/standard apps	Few operator apps and interworking (CS,PS, WLAN)
Release 7	Fixed-mobile convergence	Common IMS
Release 8	New IP only radio access, LTE	**Rel-8/Rel-9 enables CS voice replacements**
Release 9	Rel-8 "leftovers" (LTE emergency call, LCS) and some improvements	
Release 10	Increment improvements and research items	Too early to say

Figure 3.14 IMS standardisation journey – where are we?

alerting tone (CAT)], policy and charging control improvements also due to introduction of new EPC, messaging interworking, overlap signalling support for legacy public switched telephone network (PSTN) access, usage of transport layer security protocol (TLS) for IMS access security (fixed access) and IMS local breakout. This release was stabilised in December 2008.

In 3GPP Release 9 is a kind of short release the primary purpose of which is to complete Release 8 leftovers, especially in LTE and EPC area. From an IMS perspective the most important features specified in Release 9 are emergency call support in LTE and EPC and LTE positioning mechanisms. Once these were completed we can say that 3GPP finally delivered all building blocks to realise what was planned in Release 2000; all-IP architecture which enables CS telephony replacement (see Figure 3.14). Release 9 was stabilised in 2010.

3.4.3 IMS Fundamentals

There is a set of basic requirements which guides the way in which the IMS architecture has been created and how it should evolve in the future. The following issues form the baseline for the IMS architecture:

- IP connectivity and roaming;
- IP multimedia sessions;
- Access independence and layered design;
- QoS and IP policy control;
- Service control;

- Interworking with other networks;
- Charging;
- Secure communication.

3.4.3.1 IP Connectivity and Roaming

As the name IMS implies, a fundamental requirement is that a device has to have IP connectivity to access it. The UE establishes an IP connectivity by performing an EPS attach. During the attach procedure the EPC assigns an IP address to the UE as is further described in Sections 3.3.4 and 5.3. This IP address can be either IPv4 or IPv6 and it is used further in an IMS registration to bind user's identity such as +358 40 1234567 to the assigned IP address, see Section 5.4 for details.

When a user is located in the home network all elements (enodeB, MME, Serving/PDN GW, IMS) are obviously in the home network and IP connectivity is obtained in that network.

In a roaming situation, 3GPP specifications define that IP connectivity can be obtained either from the home network or the visited network. In deployed GPRS networks and LTE data only networks this IP connectivity is practically always obtained from the home network which means that enodeB, MME and S-GW are located in the visited network and P-GW is located in the home network when a user is roaming in the visited network. Applying this model for VoLTE would be problematic because it would imply that the IMS entry point is also located in the home network.

- IMS entry point in the home network means that the visited network would not anymore be aware of voice calls and short messages which would lead to smaller roaming revenues for communication service providers.
- Fulfilling regulatory requirements would be challenging or impossible.
  - IMS signalling traffic is expected to be encrypted by Internet protocol security (IPSec). This means that the visited communication service provider cannot properly perform a lawful interception.
  - IMS emergency session handling requires the use of proxy call session control function (P-CSCF) in the visited network.

In addition, P-GW in the home network increases packet latency in the user plane. Consider routing real-time transport protocol (RTP) voice packets from the USA to Europe and then back to the USA due to the use of home network PDN GW. Based on these reasons, the global system for mobile communication association (GSMA) decided in 2010 to require use of visited P-GW and IMS entry point for roaming VoLTE users. This agreement is captured in a GSMA permanent reference document (IR.65).

3.4.3.2 IP Multimedia Sessions

Existing communication networks are able to offer voice, video and messaging type of services using CS bearers. Naturally, the end users' service offerings should not decline when users move to the PS domain and start using the IMS. The IMS will take communication to the next level by offering enriched communication means. IMS users are able to mix and match a variety of IP-based services in any way they choose during a single communication session. Users can integrate voice, video and text, content sharing and presence as part of their communication and can add or drop services as and when they choose. For example two people can start a session as a voice session and later on add a game or video component to the same session.

3.4.3.3 Access Independence and Layered Design

Although this is a book about VoLTE it is important to realise that the IMS is designed to be access independent so that the IMS services can be provided over any IP connectivity networks (e.g. GPRS, WLAN, broadband access xDSL, CDMA, WiMAX). In fact the first IMS release, Release 5, is tied to UMTS as the only possible Internet protocol connectivity access network (IP-CAN) is GPRS, but in Release 6 onwards access-specific issues are separated from the core IMS description and the IMS architecture returned to its born state (i.e. access independent; see Figure 3.15). A number of different accesses have been added since then. WLAN access to the IMS was added in 3GPP Release 6, fixed broadband access was added in Release 7, DOCSIS®, cdma2000® and EPS were included in Release 8 and fibre access was added in Release 10. In addition, via additional IMS centralized services (ICS) enhanced mobile services switching centre server (eMSS) and access gateway control function (AGCF), legacy PSTN, plain old telephone service (POTS) and 2G/3G CS mobile voice subscribers can be connected to the IMS service engine.

In addition to access independence IMS architecture is based on a layered approach. This means that transport and bearer services are separated from the IMS signalling network and session management services. Further services are run on top of the IMS signalling network. The layered approach aims at a minimum dependency between layers. The layered approach increases the importance of the application layer. When applications

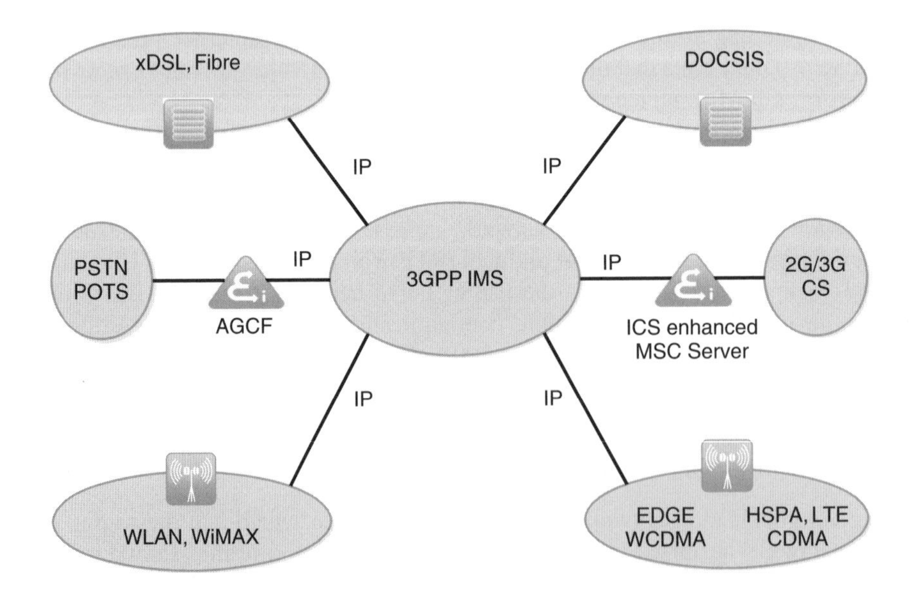

Figure 3.15 One common IMS for all accesses.

are isolated and common functionalities can be provided by the underlying IMS network the same applications can run on UE using diverse access types. Different services have different requirements. This means that in order for different services to be executed properly, the network has to be equipped with access-aware control and service control for multimedia services.

3.4.3.4 Quality of Service and IP Policy Control

On the public Internet, delays tend to be high and variable, packets arrive out of order and some packets are lost or discarded. This cannot be the case with the VoLTE. The underlying access and transport networks together with the IMS will provide end to end QoS. Via the IMS, the UE negotiates its capabilities and expresses its QoS requirements during a SIP session setup or session modification procedure.

The UE is able to negotiate such parameters as: media type, direction of traffic, media type bit rate, packet size, packet transport frequency, usage of RTP payload for media types and bandwidth adaptation. This type of information is delivered to PCRF by IMS when session is being established. The PCRF creates policy rules based on session data and push appropriate policy rules (including bandwidth, required quality class, IP packet filters) to P-GW which interprets the rules and takes actions to establish required EPC bearers for IMS applications such voice. With the help of PCRF the IMS is able to control the usage of bearer traffic intended for the IMS media. This requires interaction between the IP CAN and the IMS. The means of setting up interaction can be divided into three different categories (3GPP TS 22.228; 3GPP TS 23.203; 3GPP TS 23.228):

- The policy control element maps values negotiated in SIP signalling to policy and charging control rules for media traffic. This allows a communication service provider to utilise bearer resources optimally for SIP session. See also Section 5.5.4.
- The policy control element is able to control when media traffic between the end points of a SIP session start or stop. This makes it possible to prevent the use of the bearer resources until session establishment is completed and allows traffic to start/stop in synchronisation with the start/stop of charging for a session in IMS. See also Section 5.5.4 and 5.5.6.
- The policy control element is able to receive notifications when the IP CAN service has either modified, suspended or released the bearer(s) of a user associated with a session. This allows IMS to release an ongoing session because, for instance, the user is no longer in the coverage area. See also Section 5.5.6.2.

3.4.3.5 Service Control

In CS mobile networks the visited service control is in use. This means that, when a user is roaming, an entity in the visited network provides services and controls the traffic for the user. This entity is called a visited mobile service switching centre. In the early days of Release 5 both visited and home service control models were supported. Supporting two models would have required that every problem have more than one solution; moreover, it would reduce the number of optimal architecture solutions, as simple solutions may not fit both models. Supporting both models would have meant additional extensions for Internet engineering task force (IETF) protocols and increased the work involved in registration and session flows. The visited service control was dropped because it was a complex solution and did not provide any noticeable added value compared with the home service control. On the contrary, the visited service control imposes some limitations. It requires a multiple relationship and roaming models between communication service providers. Service development is slower as both the visited and home network would need to support similar services, otherwise roaming users would experience service degradations. In addition, the number of intercommunication service provider reference points increase, which requires complicated solutions (e.g. in terms of security and charging). Therefore, home service control was selected; this means that the entity that has access to the subscriber database and interacts directly with service platforms is always located at the user's home network. IMS service provisioning is described in more detail in Section 4.5.

3.4.3.6 Interworking with Other Networks

It is evident that the IMS is not deployed over the world at the same time. Moreover, people may not be able to switch terminals or subscriptions very rapidly. This will raise the issue of being able to reach people regardless of what kind of terminals they have or where they live. To be a new, successful communication network technology and architecture the IMS has to be able to connect to as many users as possible. Therefore, the IMS supports communication with PSTN, ISDN, mobile and Internet users. Additionally, it will be possible to support sessions with Internet applications that have been developed outside the 3GPP community (3GPP TS 22.228). Voice, video and SMS IMS-CS interworking functions are further described in Section 3.4.4.3.

3.4.3.7 Charging

From a communication service provider perspective the ability to charge users is a must in any network. The IMS architecture allows different charging models to be used. This includes, say, the capability to charge just the calling party or to charge both the calling party and the called party based on used resources in the transport level. As IMS sessions may include multiple media components (e.g. audio, video), it is required that the IMS provides a means for charging per media component. This would allow charging the called party if they add a new media component in a session. It is also required that different IMS networks are able to exchange information on the charging to be applied to a current session (3GPP TS 22.101; 3GPP TR 23.815).

The IMS architecture supports both online and offline charging capabilities. Online charging is a charging process in which the charging information can affect in real time the service rendered and, therefore, directly interacts with session/service control. In practice, an communication service provider could check the user's account before allowing the user to engage a session and to stop a session when all credits are consumed. Prepaid services are applications that need online charging capabilities. Offline charging is a charging process in which the charging information does not affect in real time the service rendered. This is the traditional model in which the charging information is collected over a particular period and, at the end of the period, the communication service provider posts a bill to the customer.

The expected commercial charging arrangements in VoLTE are:

- Zero rate charging on EPC for all IMS signalling and IMS user plane traffic;
- IMS entities collect charging information from SIP signalling and create required charging data records (CDRs);
- PCRF is used to authorise IMS bearers and control graceful release of IMS bearer and to bridge control plane and user plane error cases.

VoLTE charging arrangements are described in more detail in Section 5.5.5.

3.4.3.8 Secure Communication

Security is a fundamental requirement in every part of the mobile network the IMS is not an exception. The IMS security architecture consists of three building blocks, as illustrated in Figure 3.16. The first building block is network domain security (NDS) (3GPP TS 33.210), which provides IP security between different IMS communication service providers and entities within a single IMS communication service providers. Layered alongside NDS is IMS access security (3GPP TS 33.203). The access security for SIP-based services is a self-sustaining component in itself, with the exception that the security parameters for it are derived from the UMTS authentication and key agreement (AKA) protocol (3GPP TS 33.102). AKA is also used for bootstrapping purposes – namely, keys and certificates are derived from AKA credentials and subsequently used for securing applications that run on the hypertext transfer (or transport) protocol (HTTP) (RFC2616), among other things – in what is called the generic authentication architecture (GAA) (3GPP TS 33.220). Intentionally left out of this architectural model are those security layers that potentially lie on top of the IMS access security or run below the NDS. For

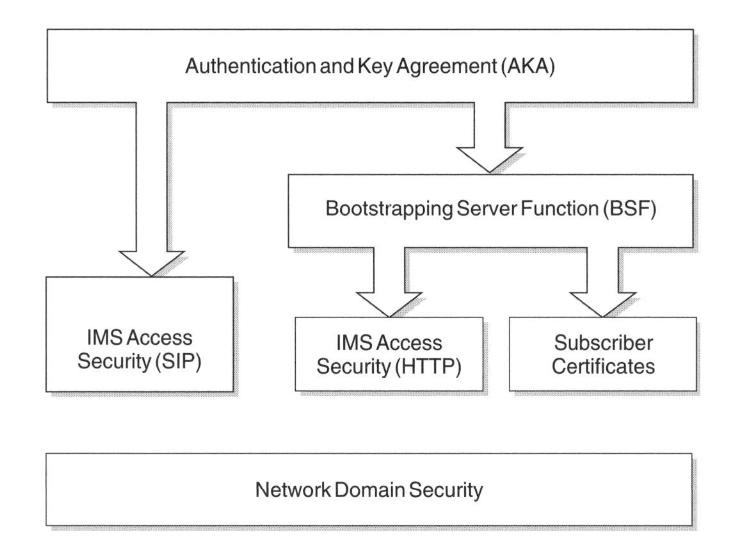

Figure 3.16 Security architecture of the IMS.

example in the LTE radio access layer implements its own set of security features, including ciphering and message integrity. However, the IMS is designed in a way that does not depend on the existence of either access security or user-plane security.

3.4.4 IMS Entities

This section discusses IMS entities and key functionalities. These entities can be roughly classified into five main categories:

- Session management and routing family (CSCFs);
- Services [AS, multimedia resource function controller (MRFC), multimedia resource function processor (MRFP), media resource broker (MRB)];
- Interworking functions [breakout gateway control function (BGCF), media gateway control function (MGCF), IMS-MGW, SGW];
- Support functions [security gateway (SEG), IBCF, transition gateway (TrGW), location retrieval function (LRF), IMS access gateway];
- Charging.

3.4.4.1 Call Session Control Functions

There are four different kinds of call session control function (CSCF): (i) proxy call session control function (P-CSCF), (ii) serving call session control function (S-CSCF), (iii) interrogating call session control function (I-CSCF) and (iv) emergency call session control function (E-CSCF). Figure 3.17 shows how CSCFs are connected to each other and other IMS entities. Each CSCF has its own special tasks and these tasks are highlighted in this section. Common to P-CSCF, S-CSCF and I-CSCF is that they all play a role during registration and session establishment and form the SIP routing machinery.

Moreover, all functions are able to send charging data to an offline charging function. There are some common functions that P-CSCF and S-CSCF are able to perform. Both entities are able to release sessions on behalf of the user (e.g. when S-CSCF detects a hanging session or P-CSCF receives a notification that a media bearer is lost) and are able to check that the content of the SIP request or response conforms the communication service provider's policy and user's subscription (e.g. content of the SDP payload contains media types or codecs which are allowed for a user).

Proxy Call Session Control Function: P-CSCF is the first contact point for users within the IMS. It means that all SIP signalling traffic from the UE will be sent to the P-CSCF. Similarly, all terminating SIP signalling from the network is sent from the P-CSCF to the UE. There are five unique tasks assigned for the P-CSCF: SIP compression, IPSec security association, interaction with PCRF, control of network address translator (NAT) and emergency session detection.

As the SIP protocol is a text-based signalling protocol, it contains a large number of headers and header parameters, including extensions and security-related information which means that their message sizes are larger than with binary-encoded protocols. For speeding up the session establishment 3GPP has mandated the support of SIP compression between the UE and P-CSCF in all other 3GPP accesses than in LTE. The P-CSCF needs

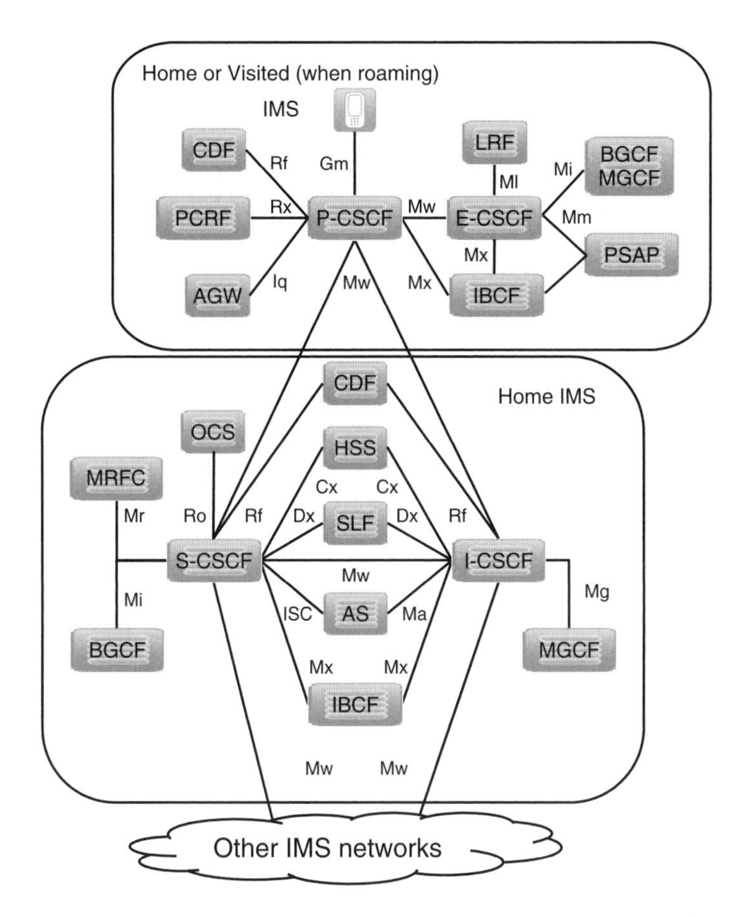

Figure 3.17 Call session control functions and reference points.[1]

to compress messages if the UE has indicated that it wants to receive signalling messages compressed.

P-CSCF is responsible for applying integrity and confidential protection for SIP signalling using IPSec or TLS. This is achieved during SIP registration. After the initial registration the P-CSCF is able to apply integrity and confidential protection of SIP signalling.

The P-CSCF is tasked to relay session and media-related information to the PCRF when an communication service provider wants to apply policy and charging control. Based on the received information the PCRF is able to derive authorised IP QoS information and charging rules that will be passed to the access gateway (e.g. PDN GW). Moreover, via the PCRF and P-CSCF the IMS is able to deliver IMS charging correlation information to the access network and, similarly, via the PCRF and P-CSCF the IMS is able to receive access charging correlation information from the access network. This makes

[1] This figure does not show reference points to AGCF, eMSS and access transfer control function (ATCF) (see Figure 3.21).

it possible to merge CDRs coming from the IMS and access networks in the billing system.

In fixed access UE is typically behind NAT which modifies the IP addresses and port information of all IP packets that traverse it. The problem arises because typically this type of NAT does not take into account IP address information in SIP and SDP layer. To make UE truly globally reachable P-CSCF is required to ensure that IP address information both in SIP/SDP and in user plane are from public IP address space. To modify the IP address in the user plane the P-CSCF controls the IMS access gateway which takes care of modifying the IP addresses in the user plane.

P-CSCF plays an important role in IMS emergency session handling as the P-CSCF is tasked to detect emergency requests in all possible cases and to select E-CSCF to handle the emergency session further.

Interrogating Call Session Control Function: I-CSCF is a contact point within an communication service provider's network for all connections destined to a subscriber of that network communication service provider. There are three unique tasks assigned for the I-CSCF:

- Obtaining the name of the next hop (either S-CSCF or AS) from the HSS.
- Assigning an S-CSCF based on received capabilities from the HSS. The assignment of the S-CSCF will take place when a user is registering with the network or a user receives a SIP request while they are unregistered from the network but has services related to an unregistered state (e.g. voice mail).
- Routing incoming requests further to an assigned S-CSCF or the AS (in the case of public service identity).

Serving Call Session Control Function: S-CSCF is the focal point of the IMS as it is responsible for handling registration processes, making routing decisions and maintaining session states and storing the service profile(s). When a user sends a registration request it will be routed to the S-CSCF, which downloads authentication data from the HSS. Based on the authentication data it generates a challenge to the UE. After receiving the response and verifying it the S-CSCF accepts the registration and starts supervising the registration status. After this procedure the user is able to initiate and receive IMS services. Moreover, the S-CSCF downloads a service profile from the HSS as part of the registration process and delivers user and device specific information to the registered UE. A service profile is a collection of user-specific information that is permanently stored in the HSS. The S-CSCF downloads the service profile associated with a particular public user identity (e.g. joe.doe@ims.example.com) when this particular public user identity is registered in the IMS. The S-CSCF uses information included in the service profile to decide when and, in particular, which AS(s) is(are) contacted when a user sends a SIP request or receives a request from somebody. Moreover, the service profile may contain further instructions about what kind of media policy the S-CSCF needs to apply – for example it may indicate that a user is only allowed to use audio and application media components but not video media components. The S-CSCF is responsible for key routing decisions as it receives all UE-originated and UE-terminated sessions and transactions. When the S-CSCF receives a UE-originating request via the P-CSCF it needs to decide if ASs are contacted prior to sending the request further on. After possible AS interaction(s) the S-CSCF either continues a session in IMS or breaks to other domains (CS or another IP network). When the UE uses a MSISDN number to address a called party then the S-CSCF converts the MSISDN number [i.e. a tel Universal Resource Locator (URL)] to SIP URI format prior to sending the request further, as the IMS does not route requests based on MSISDN numbers. Similarly, the S-CSCF receives all requests which will be terminated at the UE. Although, the S-CSCF knows the IP address of the UE from the registration it routes all requests via the P-CSCF, as the P-CSCF takes care of, for example access security functions. Prior to sending a request to the P-CSCF, the S-CSCF may route the request to an AS(s), for instance, checking possible redirection instructions. In addition, the S-CSCF is able to send accounting-related information to the OCS for online charging purposes (i.e. supporting pre-paid subscribers).

Emergency Call Session Control Function: E-CSCF is a dedicated functionality to handle IMS emergency requests such as sessions towards police, fire brigade and ambulance. The main task of E-CSCF is to select an emergency centre also known as a public safety answering point (PSAP) where an emergency request should be delivered. Typically a selection criterion is a calling user's location and possible type of emergency (e.g. police, coast guard). Once the appropriate emergency centre is selected the E-CSCF routes the request to the emergency centre.

3.4.4.2 Service Functions

Four functions in this book are categorised as IMS service-related functions – namely MRFC, MRFP, MRB and AS. Figure 3.18 shows how service functions are connected to each other and other IMS entities.

Strictly speaking, ASs are not pure IMS entities; rather, they are functions on top of IMS. However, ASs are described here as part of IMS functions because ASs are entities that provide value-added multimedia services in the IMS, such as presence and Supplementary Services. An AS resides in the user's home network or in a third-party

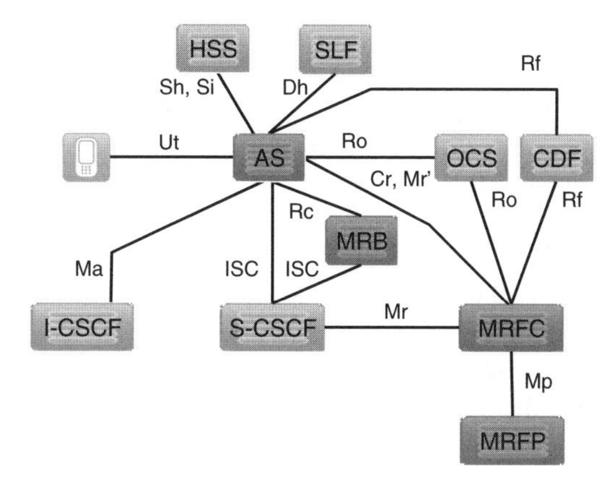

Figure 3.18 Service functions and reference points.

location. The third party here means a network or a standalone AS. The main functions of the AS are:

- The possibility to process and impact an incoming SIP session received from the IMS;
- The capability to originate SIP requests;
- The capability to send accounting information to the charging functions.

The services offered are not limited purely to SIP-based services since an communication service provider is able to offer access to services based on the customised applications for mobile network enhanced logic (CAMEL) service environment (CSE) and the open service architecture (OSA) for its IMS subscribers (3GPP TS 23.228). Therefore, AS is the term used generically to capture the behaviour of the SIP AS, OSA service capability server (SCS) and CAMEL IP multimedia service switching function (IM-SSF). From the perspective of the S-CSCF SIP AS, the OSA SCS and the IM-SSF exhibit the same reference point behaviour. An AS may be dedicated to a single service and a user may have more than one service, therefore there may be one or more AS per subscriber. Additionally, there may be one or more AS involved in a single session. For example an communication service provider could have one AS to control voice supplementary services (e.g. redirecting all incoming voice sessions to an answer machine between 5 p.m. and 7 a.m.) and another AS to handle voice call continuity (e.g. ensuring seamless handover from LTE VoIP to 2G CS call).

MRFC and MRFP together provide mechanisms for bearer-related services such as conferencing, announcements to a user or bearer transcoding in the IMS architecture. The MRFC is tasked to handle SIP communication to and from the S-CSCF/AS and to control the MRFP. The MRFP in turn provides user-plane resources that are requested and instructed by the MRFC. The MRFP performs the following functions:

- Mixing of incoming media streams (e.g. for multiple parties);
- Media stream source (for multimedia announcements);
- Media stream processing (e.g. audio transcoding, media analysis) (3GPP TS 23.002; 3GPP TS 23.228).

MRB supports the sharing of a pool of heterogeneous multimedia resource function (MRF) resources by multiple heterogeneous applications. The MRB assigns (and later releases) specific suitable MRF resources to calls when being addressed by S-CSCF or AS.

3.4.4.3 IMS-CS Interworking Functions

This section introduces five interworking functions, which are needed to enable voice, video and SMS interworking between IMS and the CS CN.

Voice and Video Interworking: Section 3.4.4.1 explained that the S-CSCF decides when to break voice or video session to the CS CN. For breaking out the S-CSCF sends a SIP session request to the BGCF; it further chooses where a breakout to the CS domain occurs. The outcome of a selection process can be either a breakout in the same network in which the BGCF is located or another network. If the breakout happens in the same network, then the BGCF selects a MGCF to handle the session further as depicted in

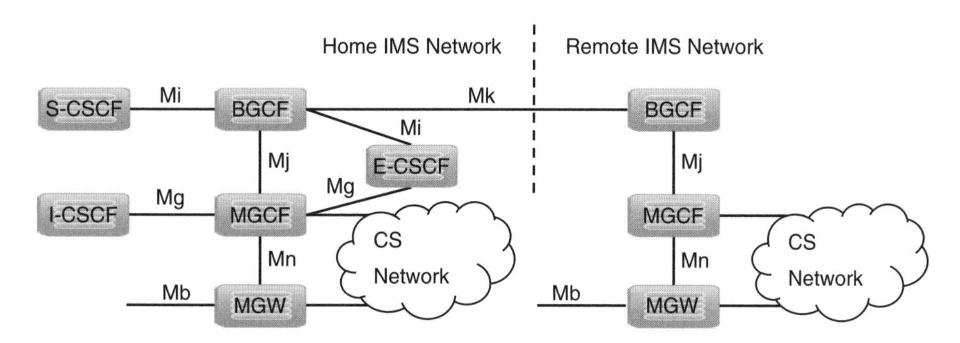

Figure 3.19 IMS-CS voice interworking functions and reference points.

Figure 3.19. If the breakout takes place in another network, then the BGCF forwards the session to another BGCF in a selected network (3GPP TS 23.228) as shown in Figure 3.19. The latter option allows routing of signalling and media over IP near to the called user.

When a SIP session request hits the MGCF it performs protocol conversion between SIP protocols and the ISDN user part (ISUP), or the bearer independent call control (BICC) and sends a converted request via the Signalling Gateway SGW to the CS CN. The SGW performs signalling conversion (both ways) at the transport level between the IP-based transport of signalling [i.e. between Sigtran SCTP/IP and SS7 message transfer part (MTP)] and the signalling system number 7 (SS7) based transport of signalling. The SGW does not interpret application layer (e.g. BICC, ISUP) messages. The MGCF also controls the IMS-MGW. The IMS-MGW provides the user-plane link between CS CN networks and the IMS. It terminates the bearer channels from the CS network and media streams from the backbone network [e.g. RTP streams in an IP network or ATM adaptation layer/asynchronous transfer mode (AAL2/ATM) connections in an ATM backbone], executes the conversion between these terminations and performs transcoding and signal processing for the user plane when needed. In addition, the IMS-MGW is able to provide tones and announcements to CS users. Similarly, all incoming call control signalling from a CS user to an IMS user is destined to the MGCF that performs the necessary protocol conversion and sends a SIP session request to the I-CSCF for a session termination. At the same time, the MGCF interacts with the IMS-MGW and reserves necessary IMS-MGW resources at the user plane.

SMS Interworking: The IP short message gateway (IP-SM-GW) is the entity that bridges the most deployed mobile messaging technology, SMS, to the IMS messaging solution. When SMS is sent to an IMS user the SMS gets routed to the IP-SM-GW which attaches the actual SMS as a special content type to the SIP MESSAGE method and passes the created SIP message to the S-CSCF for delivery. This would enable delivering SMS to devices that are attached to non-cellular IP CANs (e.g. WLAN, WiMAX). It could be also seen as an alternative termination bearer to current SMS bearer options (CS, GPRS). Utilising this type of interworking all kinds of existing value added SMS services can be delivered to users connected to the IMS. The IP-SM-GW naturally provides conversion to the other direction as well, when an IMS user is sending a SIP message containing the SMS as a special content type the IP-SM-GW extracts the actual SMS and passes it to the SMS centre for ordinary SMS delivery. This overall functionality is called SMS over

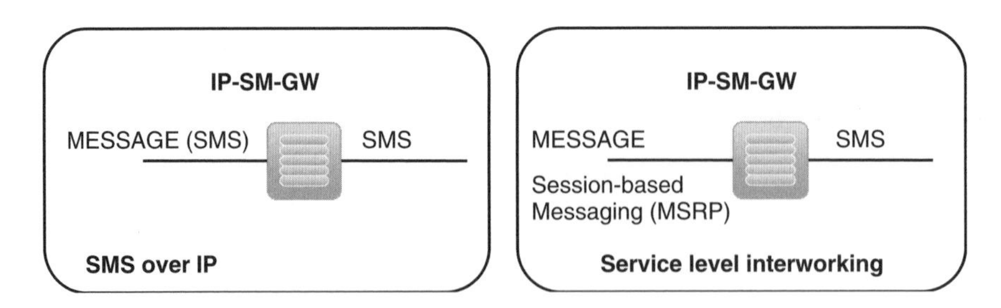

Figure 3.20 Messaging interworking.

IP in (3GPP TS 23.204); see also Figure 3.20. For a more detailed functional description see Section 5.9.

In addition the IP-SM-GW is able to provide native service level interworking between SMS and SIP-based messaging solutions. It means that SMS is fully converted to SIP-based request meaning that IMS UE (e.g. PC) does not need to implement SMS. This is depicted in the right-most part of Figure 3.20. There is one additional factor which should be taken into account in IMS messaging interworking: the size of the message. The SIP message (RFC3428) contains a size limitation. The requests for comments (RFC) states that it is not allowed to send a SIP message outside of session if the message size is at least 200 bytes less than the lowest maximum transmission unit (MTU) value found en route (usually this limit is 1300 bytes). If IP-SM-GW receives a concatenated SMS message (a group of messages formed of several standard length short messages to be sent together as if they were one longer message) and the size limit of the SIP MESSAGE would be exceeded then the IP-SM-GW should use session mode messaging to send the message. Session mode messaging means the IP-SM-GW sends a SIP INVITE request to setup a session between IMS UE and IP-SM-GW. Once the session is established the message session relay protocol (MSRP) is used to transfer the received messages to the IMS UE.

3.4.4.4 Support Functions

Several functions in this book are categorised as support functions – namely, IBCF, TrGW, IMS access gateway, SEG and LRF. For completeness we also introduce two functions which are required to connect CS, PSTN and ISDN network users to the IMS, eMSS and AGCF. Figure 3.21 shows how support functions are connected to other IMS entities (SEG is not shown).

Interconnection Border Control Function and Transition Gateway: An IBCF provides specific functions in order to perform interconnection between two communication service provider domains. It enables communication between IPv6 and IPv4 IMS applications, transcoding between different media codecs, network topology hiding, controlling transport plane functions, screening of SIP signalling information, selecting the appropriate signalling interconnect and generation of CDRs.

Capability to translate between IPv4 and IPv6 addresses emerges when IMS communication takes place between communication service providers that are supporting different IP address versions. IBCF is tasked to bridge these two domains by acting as an application

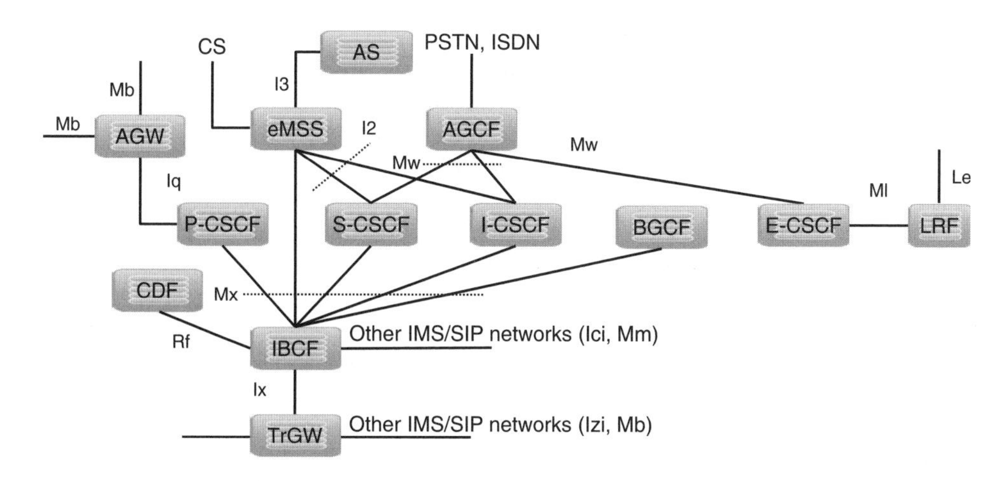

Figure 3.21 IMS support functions and reference points.

level gateway (ALG). ALG takes care of modifying SIP and SDP information in such a way that UEs using different (IPv6, IPv4) IP version can communicate with each other. The ALG functionality inside the IBCF controls TrGW.

The IBCF/TrGw provides the necessary function for codec transcoding when end-user devices that do not share a common codec, for example because they connect to different access networks. Transcoding services can be triggered proactively (before the session request is sent to the called UE) or reactively (after the session request has been sent to, and rejected by, the called UE) (3GPP TS 23.228). As an example IBCF/TrGw could provide transcoding between adaptive multi-rate (AMR) and G.722 speech codecs.

Network topology hiding functionality could be used to hide the configuration, capacity and topology of the network from outside an communication service provider's network. If an communication service provider wants to use hiding functionality then the communication service provider must place an IBCF in the routing path when receiving requests or responses from other IMS networks. Similarly, the IBCF must be placed in the routing path when sending requests or responses to other IMS networks.

The IBCF performs the encryption and decryption of all headers which reveal topology information about the communication service provider's IMS network.

A communication service provider may use IBCF to support its local policy. An communication service provider may use IBCF as the entry/exit point for its network and IBCF can be used to screen signalling information (i.e. omit or modify some received SIP headers prior to forwarding SIP messages further to other networks).

TrGW provides functions for NAT/port translation and IPv4/IPv6 protocol translation in the transport plane (i.e. modifying IP packets transporting actual IMS application media such as RTP).

IMS Access Gateway: In fixed access deployments the UE can be located behind a so-called customer premise equipment (CPE) that is the entry point to the customer's private network (this is not expected to be the case with 3GPP mobile accesses). A set top box terminating DSL in the customer's home is an example of such a CPE. A CPE acts as a NAT and firewall and modifies the IP addresses and port information of all IP packets that traverse it. The problem arises because typically this type of NAT does not take into account IP address information in SIP and SDP layer. If the UE does not take care of NAT traversal itself the IMS network needs to take care of network address translation. This capability is provided at the IMS network edge by P-CSCF which contains a SIP ALG which in turn controls the IMS access gateway (AGW) that contains a NAT.

If we assume that a SIP INVITE request is sent by the UE's private address, the ALG in P-CSCF assigns a public address and binds it to the session (it is in fact more likely that the binding stays for the time of the UEs registration, but this is left out of scope here). Within the INVITE request sent from the P-CSCF towards the terminating side, the UE's public address will be included on all protocol layers, for example:

- Within the IP packet that transports the SIP INVITE request;
- Within the SIP routing related information;
- Within the SDP message, which is conveyed in the body of the SIP INVITE request.

The ALG informs the NAT in the IMS access gateway about the newly created binding. Once media is flowing between the two UEs, the NAT in the IMS access gateway now will translate the IP addresses indicated in the RTP packets to and from the users' public and private addresses (3GPP TS 23.228; 3GPP TS 23.334; 3GPP TS 29.334).

Security Gateway: The SEG has the function of protecting control-plane traffic between security domains. The security domain refers to a network that is managed by a single administrative authority. Typically, this coincides with communication service provider borders. The SEG is placed at the border of the security domain and it enforces the security policy of a security domain towards other SEGs in the destination security domain. In the IMS all traffic within the IMS is routed via SEGs, especially when the traffic is inter-domain, meaning that it originates from a security domain which is different from the one where it is received.

When protecting interdomain IMS traffic, both confidentiality as well as data integrity and authentication are mandated (3GPP TS 33.203).

Location Retrieval Function: The LRF assists E-CSCF in handling IMS emergency sessions by delivering location information of the UE that has initiated an IMS emergency session and/or address of PSAP where the session should be sent. To provide location information the LRF may contain location server or have interface towards external location server (e.g. gateway mobile location centre). To resolve appropriate PSAP it may contain routing determination function (RDF) which is used to map the user's location to address of PSAP. The LRF may provide other emergency session parameters according to local regulations, for example this information may include emergency service query key, emergency service routing number, last routing option in North America, location number in EU, PSAP SIP URI or Tel URI. IMS emergency sessions are described in more detail in Section 5.7.

Enhanced MSC Server: eMSS acts as a contact point for users in the CS domain (normal MSS) and at the same time it appears as a P-CSCF to the other CSCFs so it performs functions normally assigned to a P-CSCF. It means that when the UE registers (attaches) in the CS core the eMSS performs an IMS registration on behalf of UE to make UE available for IMS communication. So from the IMS point of CS UE becomes a legitimate IMS end point although it does not have itself IP connectivity. When UE makes normal CS originating mobile call the eMSS skips CS mobile originating services and converts legacy CS call control protocol to IMS session control protocol and sends an IMS voice session setup request to the IMS. The IMS provides mobile originating services and takes care of routing towards the called party. Similarly when someone calls to an user who is served by eMSS the mobile terminating call gets routed to the IMS which provides mobile terminating services and after service execution the IMS routes the mobile terminating request to the eMSS which converts IMS session control protocol to CS call control protocol and initiates CS call termination without executing mobile terminating services in the CS core.

With the help of eMSS the service provisioning is independent whether the user is CS attached, is connected via IP-CAN, or uses a legacy terminal, that is the user experience is consistent and independent from the used access network type because the services are only provided by the IMS service framework. This enables the migration of a legacy 2G/3G CS user to the IMS, as shown in Figure 3.15. In addition with the help of eMSS the telephony services can also be provided in the IMS when the VoLTE user is CS attached (e.g. while roaming in an area which does not have LTE at all). This aspect is further described in Chapter 6.

Access Gateway Control Function: AGCF is the contact point for users in PSTN/ISDN networks (analogue phones, ISDN phones). It performs the following functions: (i) controls MGWs functions located in residential and access gateways, (ii) interacts with the resource and admission control subsystem, (iii) interacts with the network attachment subsystem to retrieve line profile information and (iv) performs signalling interworking between SIP and analogue/ISDN signalling. The AGCF appears as a P-CSCF to the other CSCFs so it performs functions normally assigned to a P-CSCF on behalf of legacy terminals connected behind the MGWs (such as managing SIP registration procedures, generating asserted identities, creating charging identifiers).

3.4.4.5 Charging Functions

IMS network functions are configured to detect when a chargeable trigger condition is met. After detection, the function collects the necessary information from a SIP request and either requests permission from a charging system (online charging) to continue processing the SIP request or sends relevant information to a charging system for creating a CDR for post-processing (offline charging) and allows the SIP request to continue. A chargeable trigger could be session initiation, session modification, session termination request (session-based charging) or it could be on any SIP transaction, for example MESSAGE, PUBLISH, SUBSCRIBE requests (event-based charging). Moreover, a trigger could be the presence of some SIP header or SDP information. Based on the received information the charging system either takes credit from the user's account (online charging) or transfers CDR(s) to the billing system.

Figure 3.22 shows the high-level IMS charging architecture. The left side of the figure depicts offline charging and the right side shows online charging. From the figure you can see that all IMS entities handling SIP signalling are able to communicate with the offline charging entity [i.e. charging data function (CDF)] using a single diameter-based Rf reference point (3GPP TS 32.299). The CDF receives a diameter request also from

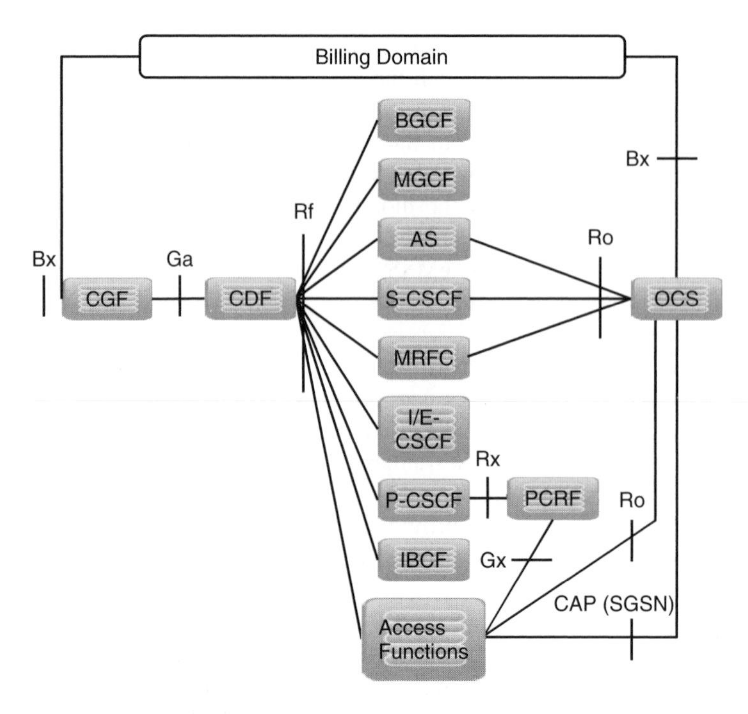

Figure 3.22 Charging interworking functions and reference points.

access network entities and based on the information provided from various entities it creates CDRs which are delivered to the charging gateway function (CGF) via the Ga reference point (3GPP TS 32.295). Finally, the CGF processes the received CDRs and transfers the final CDR(s) to the billing system using the Bx reference point (3GPP TS 32.240).

A pre-paid service requires online charging support. This means that IMS network functions need to consult the OCS before allowing users to use services. The OCS is responsible for interacting in real time with the user's account and for controlling or monitoring the charges related to service usage. Only three IMS network functions (AS, MRFC, S-CSCF) are involved in online charging. The OCS supports two reference points from non IMS network functions. SGSN uses the CAMEL application part (CAP) and the rest of the functions use the diameter-based Ro reference point. Like CGF in offline charging the OCS is also able to create CDRs in addition to credit control handling (approving resources in real time).

3.4.5 Home Subscriber Server

The HSS is the main data storage for all subscriber and service-related data. It consists of following functionalities: IMS functionality, subset of home location register and authentication centre (HLR/AUC) functionality required by the PS domain and the CS domain. HLR functionality is required to provide support to PS domain entities, such as MME. This enables subscriber access to PS domain services. In similar fashion the HLR provides

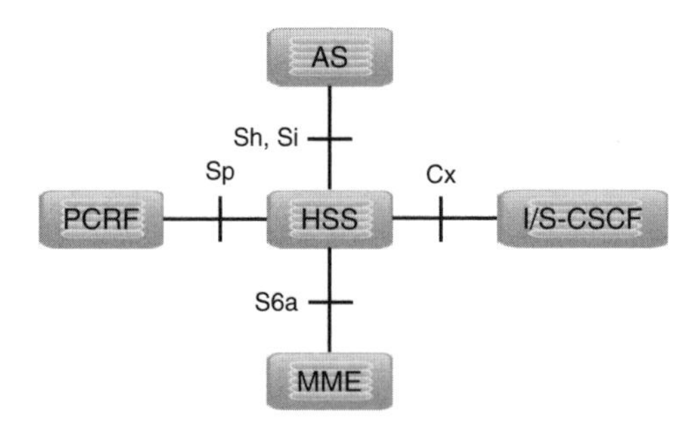

Figure 3.23 Home subscriber server and reference points.

support for CS domain entities, like MSC/MSC servers. This enables subscriber access to CS domain services and supports roaming to GSM/UMTS CS domain networks. The AUC stores a secret key for each mobile subscriber, which is used to generate dynamic security data for each mobile subscriber. Figure 3.23 shows how the HSS is connected to MME, I/S-CSCF, AS and PCRF. Please note that the Sp reference point between PCRF and HSS [or subscription profile repository (SPR)] is not standardised at the time of writing; see also Section 3.4.6.

In Section 5.2 we will provide detailed information what kind of VoLTE related data is stored in the HSS. In Section 4.3.2.1 we will explain when and how this type of information is transferred between HSS to MME; in Section 5.4.7 we will explain when and how this type of information is transferred between HSS to S-CSCF; and in Section 4.5 we will explain when and how this type of information is transferred to AS.

3.4.6 Policy and Charging Rule Function

The PCRF is responsible for making policy and charging control decisions based on session and media-related information obtained from the application function such as P-CSCF in the IMS environment. Session establishment in the IMS involves an end to end message exchange using SIP and SDP. During the message exchange UEs negotiate a set of media characteristics (e.g. codecs, IP addresses, port numbers). If an communication service provider applies the policy and charging control, then the P-CSCF will forward the relevant SDP information to the PCRF together with an indication of the originator. The PCRF generates charging rules and authorises the IP flows of the chosen media components by mapping from SDP parameters to authorised IP QoS parameters for transfer to the access network, for example P-GW. On receiving the IP bearer activation or modification, the access network function (such as P-GW) asks for authorisation information from the PCRF. Based on available information in the PCRF it makes an authorisation decision which will be enforced in the access network function. In addition to a bearer authorisation decision the PCRF receives reports on transport plane events, for example when the bearer is lost or when the bearer is released. Based on this information the PCRF is able to inform the application function (such as P-CSCF) about the occurred event. This for example allows the P-CSCF to effect charging, and it may even start releasing an IMS session on behalf of the user (see Section 5.5.6.2). Moreover, the PCRF

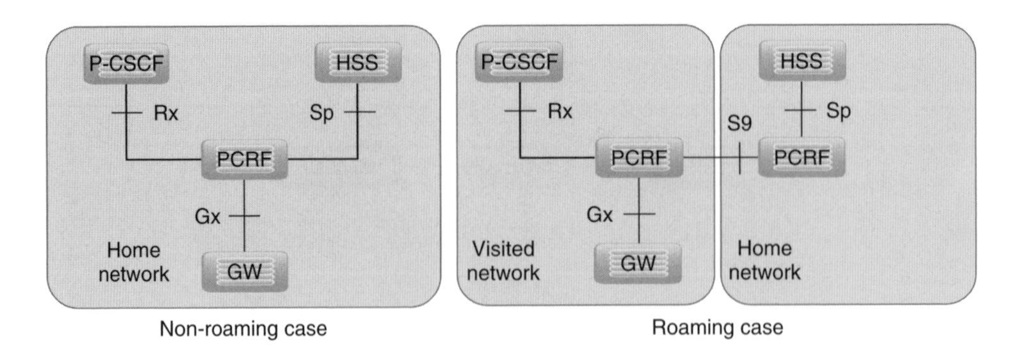

Figure 3.24 Policy and charging rule function and reference points.

can be used to exchange charging correlation identifiers which enables the communication service provider to correlate CDRs generated in the access network and IMS network. In addition PCRF can deliver following charging related information to the access network function: primary and secondary addresses of offline and online charging entity addresses, activation of online and offline charging in access gateway (enabled/disabled), metering method to be applied in access gateway (duration, volume or both), rating group information (e.g. 0.1 per minute) and desired reporting level in access gateway (based on given service or based on given service and rating-group).

Policy and charging control has experienced considerable architectural changes since its introduction in Release 5. Figure 3.24 depicts Release 10-based architecture when GTP protocol is used between gateway and access network. The SPR is logical entity contains all subscriber/subscription related information needed for subscription-based policies and IP CAN bearer level policy and charging rules by the PCRF. The Sp reference point protocol has not been specified in Release 10. In Figure 3.24 the SPR functionality is expected to be part of HSS.

Further evolution of policy and charging control is under way and foreseen in the forthcoming specification releases with the fixed-mobile convergence. The first step in this evolution has already been agreed to be the standardisation of interworking between the policy and charging control architectures of 3GPP and the Broadband Forum.

3.5 Summary

This section summarises how the previously described key network entities are connected and what protocol is used; moreover, a detailed VoLTE architecture is depicted (Figure 3.25; Table 3.3)

For the sake of clarity, it is impossible to include everything in one figure; so, please note the following:

- Figure 3.25 does not show charging related functions or reference points (see Section 3.4.4.5 and Figure 3.22 for more details).
- The figure does not show all reference point between AS and other entities (see Section 3.4.4.2 and Figure 3.18 for more details).
- The figure does not show detailed architecture for SR-VCC (see Section 5.6.2 and Figures 5.22 and 5.26).

Table 3.3 Summary of reference points

Name of reference point	Involved entities	Protocol
Bx	CGF – billing domain	File transfer protocol (FTP)
Cr	AS – MRFC	HTTP
Cx	HSS – S-CSCF/I-CSCF	Diameter
Dh	SLF – AS	Diameter
Dx	SLF – S-CSCF/I-CSCF	Diameter
Ga	CDF – CGF	Diameter
Gm	UE – P-CSCF	SIP
Gx	P-GW – PCRF	Diameter
I2	MSS – I-CSCF/S-CSCF	SIP
I3	MSS – AS	XCAP
Ici	IBCF – IBCF	SIP
Iq	P-CSCF – AGW	H.248
ISC	S-CSCF – AS	SIP
Ix	IBCF – TrGW	H.248
Izi	TrGW – TrGW	IMS user plane transport
Le	LRF – external LCS client	OMA MLP and OSA-API
LTE-Uu	UE – eNodeB	PDCP
Ma	I-CSCF – AS	SIP
Mb	P-GW, MRFP, MGW, TrGW, AS (not shown in Figure 3.1)	IMS user plane transport
Mg	MGCF – I-CSCF	SIP
Mi	S-CSCF – BGCF	SIP
Mj	BGCF – MGCF	SIP
Mk	BGCF – BGCF (not shown in Figure 3.1)	SIP
Ml	E-CSCF – LRF (not shown in Figure 3.1)	SIP
Mp	MRFC – MRFP	H.248
Mr	S-CSCF – MRFC	SIP
Mr'	AS – MRFC	SIP
Mw	CSCF – CSCF	SIP
Mx	CSCF/BGCF – IBCF	SIP
Rc	AS – MRF	SIP (not well defined)
Rf	See Figure 3.22	Diameter
Ro	See Figure 3.22	Diameter
Rx	P-CSCF – PCRF	Diameter
S11	MME – S-GW	GTP2
S1-MME	eNodeB – MME	S1-AP
S1-U	eNodeB – S-GW	GTP-U
S3	MME – SGSN	GTPv2-C
S5	S-GW – P-GW	GTPv2-C/GTP-U
S6a	MME – HSS	Diameter
S9	PCRF – PCRF	Diameter
Sh	AS – HSS	Diameter
Si	HSS – CAMEL IM-SSF	MAP
SGs	MME – MSC server	SGs AP
Sv	MSC server – MME	GTPv2-C
Ut	UE – AS	XCAP
Mm	I/S-CSCF, IBCF – external IP networks	SIP

SLF = subscription locator function; XCAP = XML configuration access protocol; AGW = access gateway; PDCP = packet data convergence protocol.

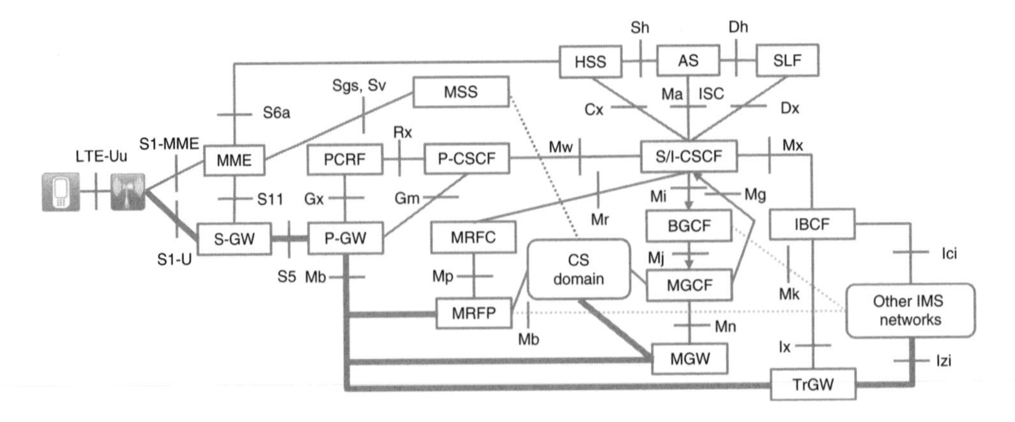

Figure 3.25 VoLTE architecture.

- The figure does not show detailed architecture for CSFB (see Section 5.8.1 and Figure 5.28).
- The figure does not show detailed architecture for IMS centralised services (ICS; see Chapter 6 and Figure 6.3).
- The figure does not show emergency session functionality.

4

VoLTE Functionality

4.1 Overview

Here we explain how voice over long-term evolution (VoLTE) user equipment (UE) obtains required Internet protocol (IP) connectivity [evolved packet core (EPC) authentication, IP address allocation, retrieval of packet data network (PDN) subscription context, default bearer activation] and discovers both that Internet protocol multimedia subsystem (IMS) based voice over Internet protocol (VoIP) is possible and what is the address of the IMS entry point. We will further explain the required radio and EPC bearers for VoLTE and associated quality of service (QoS) properties. In addition from a long term evolution (LTE) point of view, we cover packet scheduling, power saving features mobility concepts, circuit switched fallback (CSFB) options (CSFB call cases will be covered separately in Section 5.8), expected UE capabilities and UE positioning. Furthermore here we introduce two different identity modules [universal subscriber identity module (USIM) and IP multimedia services identity module (ISIM)], IMS identities, IMS service execution and IMS telephony service which are discussed further in the next chapter.

4.2 Radio Functionalities

This section describes the main radio functionalities that are needed to carry high quality voice traffic efficiently in LTE radio. First, packet scheduling and QoS differentiation are discussed, which are relevant functionalities from the voice capacity and quality perspective. Second, the mobility solutions inside LTE and from LTE to 2G and 3G radios are described. Positioning solutions are discussed, especially for emergency calls. Finally, the optional terminal features and their benefits are presented. The radio features are summarised in Figure 4.1.

4.2.1 Bearers and Scheduling

LTE uses the concept of bearers to carry the data between UE and the core network, and to provide QoS differentiation. The principle in all 3GPP radios is that all the bits within one bearer have the same QoS. If QoS differentiation is required towards a single UE, then each QoS class uses its own bearer. The default bearer in LTE carries IMS signalling by using a QoS class identifier (QCI) value of 5. The voice bearer uses a dedicated bearer

Voice over LTE: VoLTE, First Edition. Miikka Poikselkä, Harri Holma, Jukka Hongisto, Juha Kallio and Antti Toskala.

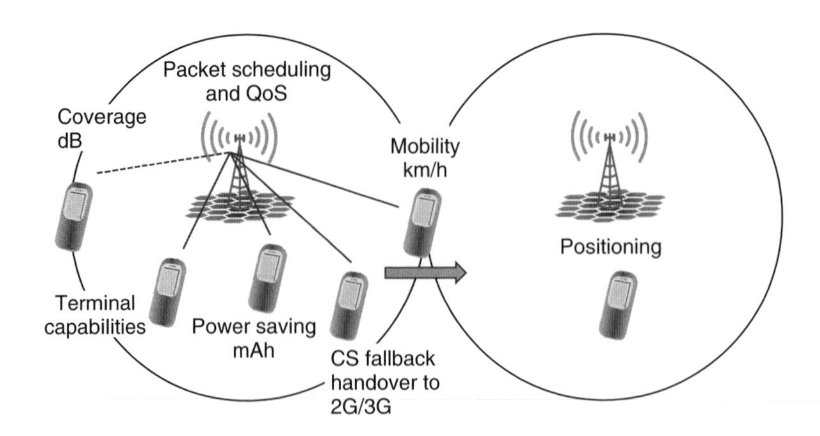

Figure 4.1 Main radio functionalities for voice support in LTE.

with a QCI of 1. There may also be further bearers for other parallel services, like file or video sharing. The bearers are shown in Figure 4.2.

There are no dedicated user-specific resources reserved in the air interface but in a typical case every single packet transmission is scheduled. The scheduling algorithms are relevant in defining the system capacity and in guaranteeing the end user quality. QoS prioritisation is provided by the packet scheduler. The principle is illustrated in Figure 4.3. The packet scheduling algorithms are not defined in 3GPP but the algorithms are left to be designed by the network vendors.

The packet scheduling for voice can use fully dynamic scheduling or semi-persistent scheduling (Figure 4.4). The dynamic solution schedules every single voice packet that arrives with 20 ms periods. The dynamic scheduling has the benefit that it gives the full freedom to select optimal radio resources for each transmission depending on the radio conditions and load conditions. The drawback is that every packet needs physical downlink control channel (PDCCH) capacity which can be the limiting factor for high traffic cases. Another approach is semi-persistent scheduling where the evolved NodeB (eNodeB) pre-allocates resources for a voice packet every 20 ms. Semi-persistent scheduling does not require control channel capacity.

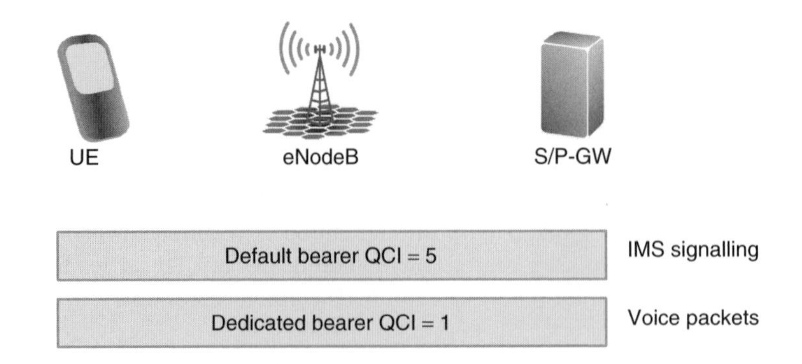

Figure 4.2 Multiple bearers for QoS differentiation.

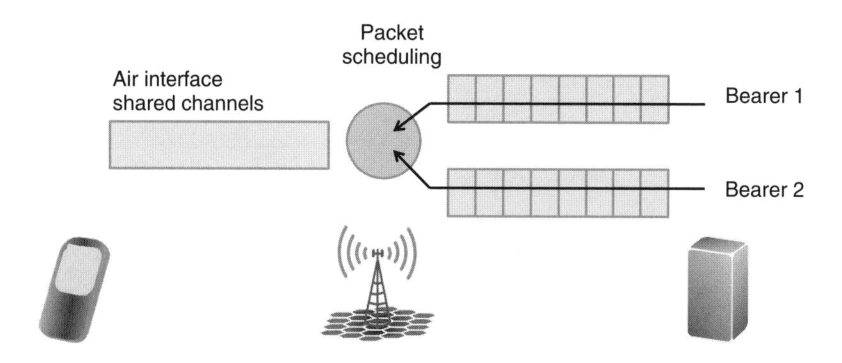

Figure 4.3 QoS differentiation provided by packet scheduling.

Another solution to improve efficiency is to combine two voice packets together before transmitting the packets in the air interface. That solution reduces the air interface overhead and improves efficiency. The combining of voice packets is called voice packet bundling. The packet bundling does not require any specific UE support but it can be implemented by the packet scheduler.

4.2.2 Mobility

LTE system is designed to support seamless mobility within LTE system and between LTE and 2G and 3G networks. The intra-frequency mobility solution in LTE is slightly different from the mobility in 2G and 3G networks because of LTE flat architecture. 2G and 3G networks use a base station controller (BSC) and radio network controller (RNC) to manage the mobility while LTE radio network uses only eNode (Base station) in the radio network architecture. The mobility management is controlled by eNodeB based on the measurements provided by the terminals.

An intra-frequency handover procedure is shown in Figure 4.5. UE is first connected to source eNodeB that has a S1 user plane connection to the core network gateway

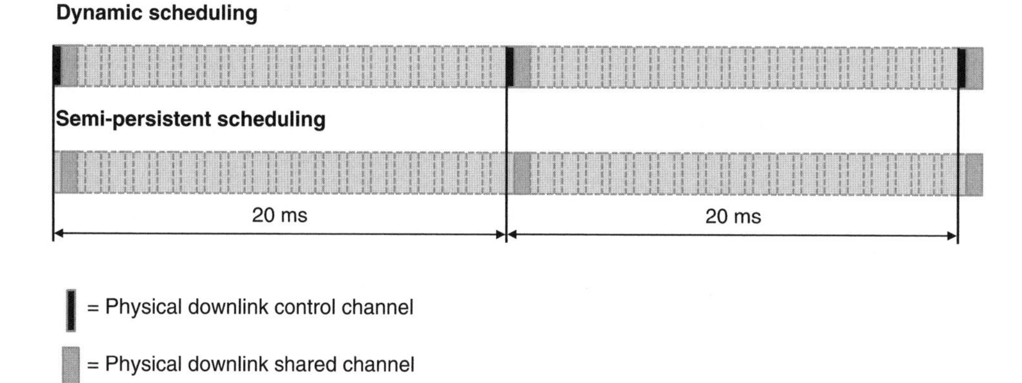

Figure 4.4 Dynamic and semi-persistent scheduling for voice.

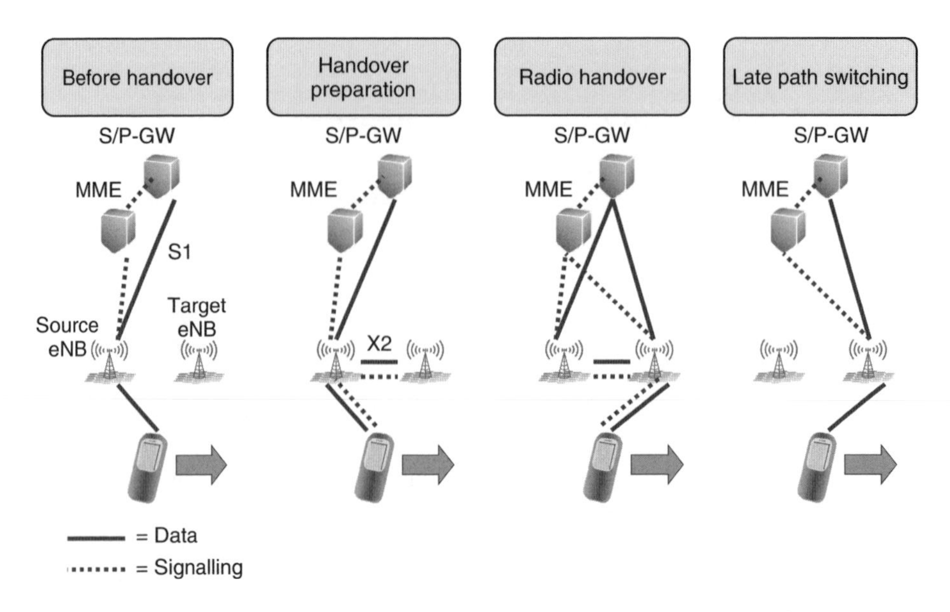

Figure 4.5 Intra-frequency mobility. Reproduced with permission from John Wiley & Sons, Ltd.

and control plane to the mobility management entity (MME). UE runs intra-frequency neighbour cell measurements continuously. When the target cell becomes strong enough, UE will send the measurement report to the source eNodeB. The measurement triggers are defined by the network. The source eNodeB reserves resources in the target cell over the X2 interface. If the X2 interface is not available, the source eNodeB can establish an X2 connection by using the measurement information from UE and the information from network management system. In the second phase, the source eNodeB sends a handover command to UE. The source eNodeB starts forwarding the downlink packets towards the target eNodeB over X2 interface. The X2 interface allows the use of lossless handovers where no packets are lost. In the third phase, UE changes the radio connection from the source eNodeB to the target eNodeB. The break in the physical connection during the handover is typically less than 25 ms. These first phases of the handover are not visible to the core network. In the last phase, the target eNodeB requests the core network to update the connection from the source eNodeB to the target eNodeB. This phase is called 'late path switching' and is not visible to the UE. The idea in this mobility is to make the radio handover as fast as possible to maintain reliable connection and only then update the core network connection. The flat architecture implies that every handover causes some signalling to the core network. In 2G/3G networks BSC and RNC hide most of the mobility from the core network. Therefore, the core network signalling capacity in LTE must be designed high enough.

LTE inter-frequency and inter-system handovers have a few differences compared to an intra-frequency handover. LTE UE can make inter-frequency and inter-system measurements only when the network has allocated measurement gaps. The activation of measurement gaps can be trigged by the UE measurements, for example indicating low signal power when running out of LTE coverage. The measurements can also be activated when there is a need to make CSFB or SR-VCC handover to 2G or 3G.

4.2.3 Circuit Switched Fallback Handover

The CSFB was intended as an interim solution in Release 8, until IMS based voice support became available in the networks, as initial LTE networks do not typically support voice. There are specified means to fall back to a CS voice service in a global system for mobile communications (GSM), universal mobile telecommunications system (UMTS) or code division multiple access (CDMA) network if available in the same coverage area. From the architecture point of view the CSFB requires the availability of the SGs interface between the MME and the mobile switching centre (MSC) server to enable it to provide CS paging to the LTE side as well as combined EPC and international mobile subscriber identity (IMSI) attachment and detachment procedures (Figure 4.6). Also short message service (SMS) is delivered via this interface.

All options for CSFB to GSM, wideband code division multiple access (WCDMA) and cdma2000 are listed in Table 4.1. The challenge created with the multiple options is that it will be difficult to ensure a 'universal' solution available in all devices at the beginning as all of the solutions are not likely to be available for testing with UEs in the first phase. 3GPP is mandating Release 8-based solutions, as shown in Table 4.1, with the exception of the packet switched (PS) handover to GSM/EDGE radio access network (GERAN) for which there is a separate UE capability, as shown in (3GPP TS 36.306). Release 9 added further optimisation to address specific deployment scenarios. There were no further enhancements in Release 10.

4.2.3.1 CS Fallback to UMTS

When doing the CSFB to any system, there is the challenge that call set up may take longer than in a normal case because the UE is doing fallback and only then getting connected to the system. There are implementation means to reduce this impact, as addressed in

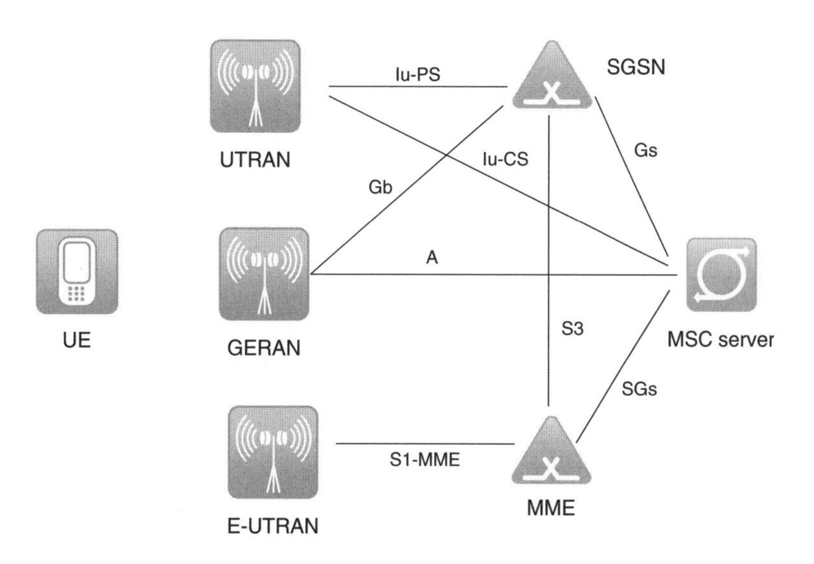

Figure 4.6 CSFB architecture.

Table 4.1 CS fallback options in Release 8 and 9

Target system	Solutions	Release	Mandatory for CS fallback UE
UTRAN	RRC connection release with redirection without Sys Info	8	Yes
	RRC connection release with redirection with Sys Info	9	No
	PS handover with data radio bearer(s) (DRB)	8	Yes
GSM	RRC connection release with redirection without Sys Info	8	Yes
	RRC connection release with redirection with Sys Info	9	No
	Cell change order without NACC	8	Yes
	Cell change order with NACC	8	Yes
	PS handover	8	No
cdma2000 (1×RTT)	RRC connection release with redirection	8	Yes
	Enhanced 1×CSFB	9	No
	Enhanced 1×CSFB with concurrent high rate packet data (HRPD) handover	9	No
	Dual receiver 1×CSFB (RRC connection release without redirection)	9	No

HRPD = High rate packet data.

connection with each of these methods, thus the resulting extra time is expected to be in the order of 1–2 s when compared to a call set up without fallback. Once solutions such as PS handover become available this difference is expected to become even smaller.

- The use of radio resource control (RRC) connection release with redirection information is straightforward and based on releasing the RRC connection. The LTE RRC connection is released if the UE starts a CS call setup request via extended service request or if the UE receives a CS paging as shown in Figure 4.7. With the connection released the UE is redirected to the UMTS system. Once in the target radio access technology (RAT) the UE searches for a suitable cell by utilising the frequency information in the RRC connection release. In a suitable cell the UE will start from IDLE and acquires necessary system information blocks to access the cell. Acquisition time for the system information block depends a lot on the implementation of system information block scheduling and the sizes of the system information blocks. The necessary system information for initial access can be received in a matter of some hundreds of milliseconds with an efficient system information block (SIB) scheduling, while with very long scheduling intervals then obviously more time is required. There is Release 9 optimisation to address such cases where SIB scheduling would cause longer delays.
- The fastest Release 8 method is the use of a PS handover, which after completion of the handover has the benefit that the UE is already connected to the system and thus does not need to spend time for the system acquisition process. This normally requires completion of the inter-RAT measurements unless a blind handover can be used. The use of a non-guaranteed bit rate (GBR) bearer does not typically require any

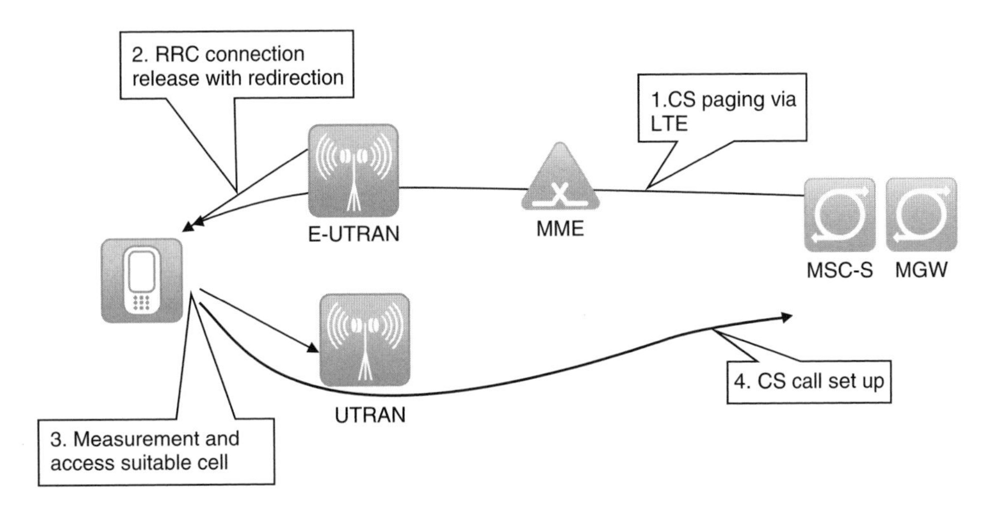

Figure 4.7 CS fall-back to UTRAN with RRC connection release with redirection information.

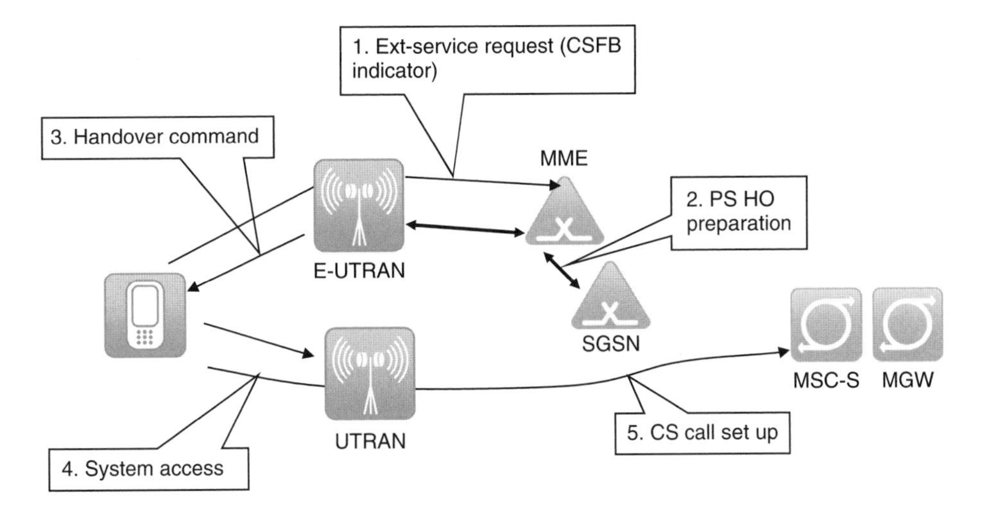

Figure 4.8 CS fall-back to UTRAN with PS handover after UE originated CS call.

unnecessary resource reservation in the target system. The UE originated CS call with PS handover to GERAN is shown in Figure 4.8.

4.2.3.2 CS Fallback to GSM

The five different solutions for doing the CSFB to GSM are partly similar to those of UMTS CSFB:

- The RRC connection release with redirection is assumed in the mandatory for to be as in Release 8 (no system info) or in Release 9 extension with system info, as shown in Figure 4.9 with CS paging as triggering.

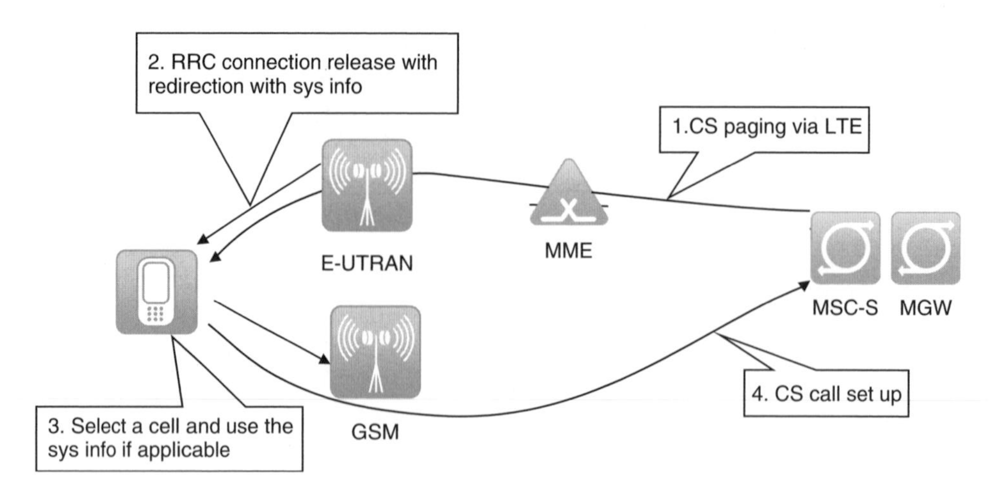

Figure 4.9 CS fall-back to GSM with RRC connection release with redirection information with sys info.

- Cell change order can be with or without network assisted cell change (NACC), both of which solutions are in Release 8. The use of NACC allows the UE to skip the system information block acquisition step.
- Then PS handover is provided as one of the options, which is not mandated to UE even though the PS handover is a Release 8 solution. This is due to the fact that not all GSM networks are expected to be updated to support the PS handover functionality, as was indicated in the discussions on the topic in 3GPP.

4.2.3.3 CS Fallback to cdma2000 System

CSFB mechanisms to the cdma2000 system, that is single carrier radio transmission technology (1×RTT) system are covering the following alternatives:

- RRC connection release with redirection to 1×RTT which is assumed to be in all devices (mandatory Rel-8 mechanism);
- Enhanced 1×CSFB where the handover signalling is tunnelled between UE and 1×RTT network. The 1×RTT channel assignment message allows the UE to acquires a traffic channel in the 1×RTT network.

Dual receiver 1×CSFB is RRC connection release without redirection information, as the UE with the dual receiver is already connected to 1×RTT and thus the UE may do the the normal 1×CS call origination or termination procedure in the 1×RTT network.

4.2.4 Mobility from 2G/3G Back to LTE

Following the move away from LTE for different reasons (lack of LTE coverage, handover due other reasons or CSFB), there is typically desire to move the UE back to LTE assuming coverage is available. This allows to utilise the most performing network as well as enables also highest data rates for PS connections. The different means available are as follows:

- With momentary loss of LTE coverage, the UE can be directed to back to LTE with the cell reselection parameters as then LTE can be considered as the highest priority system. This is not for the cell_dedicated traffic channel (DCH; connected mode) but once the connection is released or the UE in high-speed packet access (HSPA) is moved either to idle, cell_paging channel (PCH), or UTRAN registration area (URA)_PCH state, or respectively to idle mode in GSM. Similar after the voice call (with CSFB done first), once the call is over, the UE can be moved back to LTE with forced cell reselection. 3GPP has still recently addressed improvements in this area to ensure that a UE with an active PS connection does not get stuck in a HSPA network due frequent state transitions but will make measurements promptly after being moved to a new state in the WCDMA/HSPA network and can make reselection to LTE when that is the highest priority radio technology. The necessary clarifications on the UE behaviour were done in Release 10 specification in such a way that it can be easily included in a Release 8 implementation. Reselection is also available from cdma2000 networks to LTE.
- There are also a possibility such as PS handover to move the UE back to LTE, also from cdma2000 networks the handover is specified in 3GPP and 3GPP2 specifications.
- 3GPP is further developing solutions: the 'reverse single radio voice call continuity (SR-VCC)' is being looked at because that could enable a move from a CS voice call in GSM or WCDMA to a PS voice call in either LTE or alternatively over HSPA.

4.2.5 Power Saving Features

Power consumption is a major challenge in smartphones. Reduction of the power consumption of the terminal radio modem would be beneficial. The power saving in LTE is based on discontinuous transmission (DTX) and discontinuous reception (DRX) concepts. The main idea of DRX is that UE wakes up only to receive the control channel after certain periods. The DRX function has been used in RRC_idle state from the very beginning. The similar DRX function can be used also in RRC_connected state to further reduce the power consumption. The DRX period in RRC_idle is normally in the order of hundreds of milliseconds or even more than one second. The DRX period in RRC_connected state is generally less than 100 ms. The DRX concept in RRC_connected state includes long DRX and short DRX. The short DRX expects the terminal to wake up more often to receive the possible allocation information. This period is called short DRX cycle. If no data has been received by the terminal during a predefined period, the DRX short cycle timer, the terminal will move to use the long DRX cycle. The long DRX cycle allows the terminal to save more power and to use sleep mode more efficiently. The concept of long and short DRX is shown in Figure 4.10. If the connection has been inactive for longer period, like tens of seconds, then typically the RRC connection is released and the terminal is moved to RRC_idle. The terminal power consumption is lower in RRC_idle because of the longer DRX period and also because the measurement requirements are more relaxed in idle.

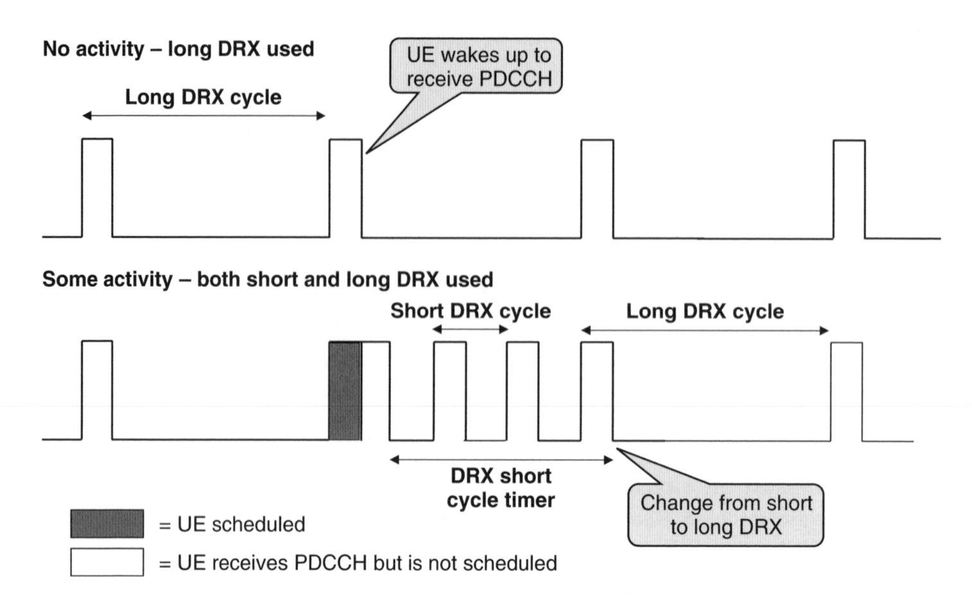

Figure 4.10 Discontinuous reception (DRX) with long and short DRX.

The DRX feature can also be used for voice calls. The short DRX can have such a short period that DRX is feasible also between the voice packets that arrive every 20 ms. LTE radio allows one to take full benefit of the DRX because the transmission time interval (TTI) is short, which leaves enough time to use efficient DRX also for voice.

4.2.6 Positioning Solutions

The original idea with LTE position location was to rely on the use of a global positioning system (GPS), either standalone or assisted GPS as the positioning solution [or assisted global navigation satellite system (A-GNSS), also covering the GLONASS satellite system], in addition to the simple cell ID method. However during the Release 9 work it was considered that especially for the needs of emergency call positioning is some markets it is necessary to have another solution as well rather than only a satellite-based alternative. For different 3GPP releases the following solutions are available in addition to satellite usage:

- Release 8 allows use of the cell ID with the server having information of the cell locations. Additionally measurements of neighbour cells can be used as well as timing advance information to provide further estimates of the UE location in the cell.
- Release 9 contains the UE measurements of the observed time difference of arrival (OTDOA) which when available from the minimum of three cells (serving cell and two others) allows one to calculate the UE location, with the principle as shown in Figure 4.11. The evolved serving mobile location center (E-SMLC) is proving UE suggested cells for measurement and then (based on the reported cell) the calculation is done on the network side, using E-SMLC which is connected to the MME. The E-SMLC

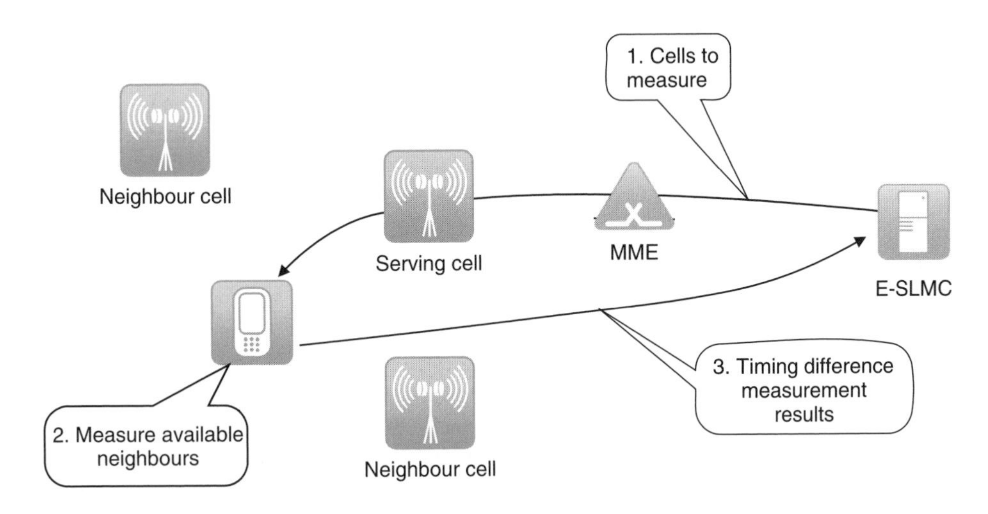

Figure 4.11 Release 9 OTDOA positioning method for UE positioning.

needs to know the location of the cells and, if the base stations are synchronised accurately, then good accuracy can be achieved.

- Release 10 does not contain new UE location solutions, while for Release 11 there is on-going work on two possible new methods. The uplink time difference of arrival (UTDOA) which aims to base the measurements on the UE uplink Sounding Reference Signal (SRS) transmission as measured in the different base station sites. The other method under investigation is the use of radio frequency (RF) pattern matching which bases the position location for the measurements done for the handover purposes in the UE. Those measurements would be then compared to earlier measurements (where the position is known) to determine the current UE position.

4.2.7 UE Radio Access Capabilities for VoLTE

There are UE capabilities which will be needed for support when the UE is capable of supporting VoLTE. As these features are mostly such that they were not needed in the first phase LTE network with data only, the approach was to create for these (and some other features) an indication mechanism with a so-called feature group indicators (FGIs) in 3GPP specifications (TS 36.331). The FGIs are intended to be set to true once testing against network has been enabled so that the network will also then have confidence that a UE indicating support for a particular feature also works. This avoids the risk with UE vendors having to roll out devices which have untested features which could then become unusable in the networks if one or more UEs would then fail to actually use the feature when activated in the network.

For the features specific to VoLTE support the following ones are considered:

- Support for radio link control (RLC) unacknowldeged mode (UM). In LTE all the point to point data is subject to the physical layer retransmissions and thus use of the RLC UM for packet reordering is needed. The first phase data only networks use the RLC acknowledged mode (AM) which enables retransmission from RLC layer if the physical layer retransmission fails.

- TTI bundling has also a separate FGI bit and there have been suggestions to mandate this (setting the support for true) for all Release 9 devices in 3GPP, but this is likely to take place only after Release 9. In early phase the TTI bundling was combined with several other features but was moved in late phase behind a separate FGI bit to make introduction easier.
- Semi persistent scheduling has also a FGI bit but so far the support has not been mandated. The feature has benefits when a large volume of data comes from voice, as shown in the results in section, but thus the support of semi persistent scheduling (SPS) was not critical for the first phase LTE networks dominated by data.
- SR-VCC related handovers to GSM, UMTS and cdma have separate FGI bits.
- The PS handover to UMTS, as discussed in Section 4.2.3, is mandatory and the related FGI bit is expected to be set to true since testing has been considered to be available (which is the requirement for 3GPP to consider mandating some FGI bit settings in a given release for a feature behind a FGI bit).
- Long DRX support allows UE battery savings and becomes important not only because of voice but also due to handheld devices with their own battery. This is assumed to be supported by all new devices currently.

For the CSFB operation, the necessary LTE measurement event behind a FGI bit from GSM, UMTS or cdma are mandated for a UE supporting CSFB for a respective technology. From the data rate capabilities point of view all LTE UE categories can support necessary data rates for voice, since the smallest Release 8, 9 and 10 data rates defined start from 10 Mbps downlink to 5 Mbps in the uplink direction, ranging up to the gigabit per second range with Release 10 LTE advanced.

4.3 EPC Functionalities

This section explains the EPC functionalities, which are essential for VoLTE. We discuss also other functionalities that are relevant when the end user has also other applications and those are used simultaneously with VoLTE.

VoLTE is based on the multimedia telephony (MMTel) service, which is the standardised IMS-based VoIP service designed to replace existing CS voice (see Section 4.6). In a circuit switched network the core domain provides both access mobility/connectivity and call control functionalities, but when VoLTE is used the EPC domain provides access mobility/connectivity, whereas IMS domain provides call control. See also IMS entities and functionalities in Section 3.3.2.

Figure 4.12 explains the basic architecture for IMS-based VoLTE.

As shown in Figure 4.12 the EPC is between LTE radio access and IMS service domain. The EPC has interfaces to mobile, radio access, subscription register, policy server and IMS service domain. Below are listed the main EPC interfaces that are needed when a VoLTE subscriber connects to EPC:

- Network access stratum (NAS) interface is used for network (L3) level signalling between UE and MME.

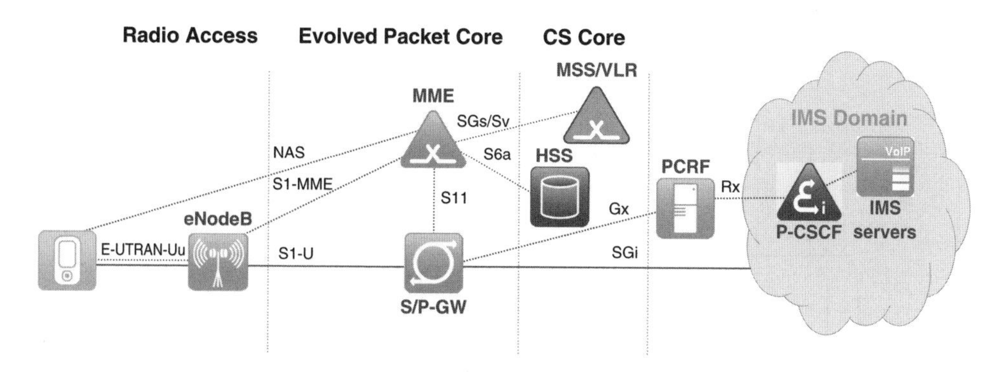

Figure 4.12 EPS architecture overview for VoLTE.

- S1-MME interface towards eNodeB is used for security, mobility and radio bearer set up.
- S1-U interface between S-GW and eNodeB is for GTP-U tunneled user plane traffic.
- S6a interface is used to obtain subscription information.
- S11 interface between MME and serving gateway (S-GW) contains all evolved packet system (EPS) bearer level functionalities, including mobility and QoS.
- SGs interface is needed for CSFB and Sv interface is used for SR-VCC.
- Gx interface between PDN gateway (P-GW) and policy and charging rules function (PCRF) is used for policy and charging control.
- SGi interface is an IP-based interface from P-GW to service domains like IMS.

See also the LTE radio architecture description in Section 3.2.2.

4.3.1 LTE subscriber identification

A USIM is an application residing on the universal integrated circuit card (UICC), which is a physically secure device that can be inserted and removed from the UE. The USIM itself stores IMSI and information for authentication subscriber identification. This data is used when a user registers his LTE device on the LTE network. The most important data stored in the USIM is listed below:

- IMSI, which is a unique identification for a mobile phone user, for example *244051234567*;
- Mobile subscriber integrated services digital network (MSISDN), which is a unique identification for subscription in mobile networks with associated services, for example *358501234567*;
- Ciphering and integrity keys for NAS security;
- Location information: for EPS location, the information consists of globally unique temporary identifier (GUTI), last visited registered tracking area identity (TAI), Evolved Cell Global Identity (E-CGI) and EPS update status;
- Emergency call information like dialling numbers.

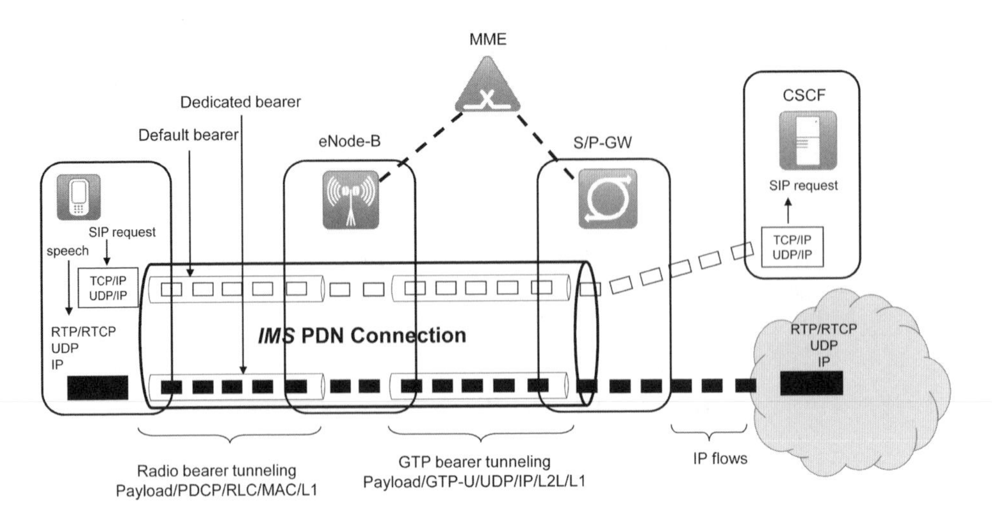

Figure 4.13 VoLTE traffic handling over EPS network.

The USIM also stores other network specific data such as SMS information, call information and public land mobile network (PLMN) information. More information is available in (3GPP TS 31.102).

4.3.2 PDN Connectivity Establishment for the VoLTE User

As described in Section 3.3 the LTE user has at least one PDN connection when the user is registered into the EPS network. For the VoLTE user the PDN connection is always established to the well known *IMS* access point name (APN). The default bearer must be created during the initial attachment for IMS signalling and the dedicated bearer for the media is created during the IMS voice call session setup. Figure 4.13 explains how session initiation protocol/XML configuration access protocol SIP signalling and real-time transport protocol/RTP control protocol (RTP/RTCP) media traffic is transferred over the LTE/EPC when a user has both IMS registration and an ongoing IMS voice call.

As shown in Figure 4.14 there are two bearers needed for VoLTE; the default bearer that is created during the EPS registration and the dedicated bearer that is created during the voice call setup. The section below further explains the EPC functionalities needed for the user registration and bearers establishment, first for IMS signalling and later for the IMS voice call. IMS procedures are explained in Section 5.

4.3.2.1 EPS Registration

When a user switches his mobile phone on, the UE makes a registration via an initial attachment to the network. In the registration the user is authenticated, authorised and the security association is created for the EPS signalling and user plane.

The user is authenticated based on the EPS authentication and key agreement (AKA) using the IMSI/P-TMSI (temporary mobile subscriber identity) and K that is a permanent

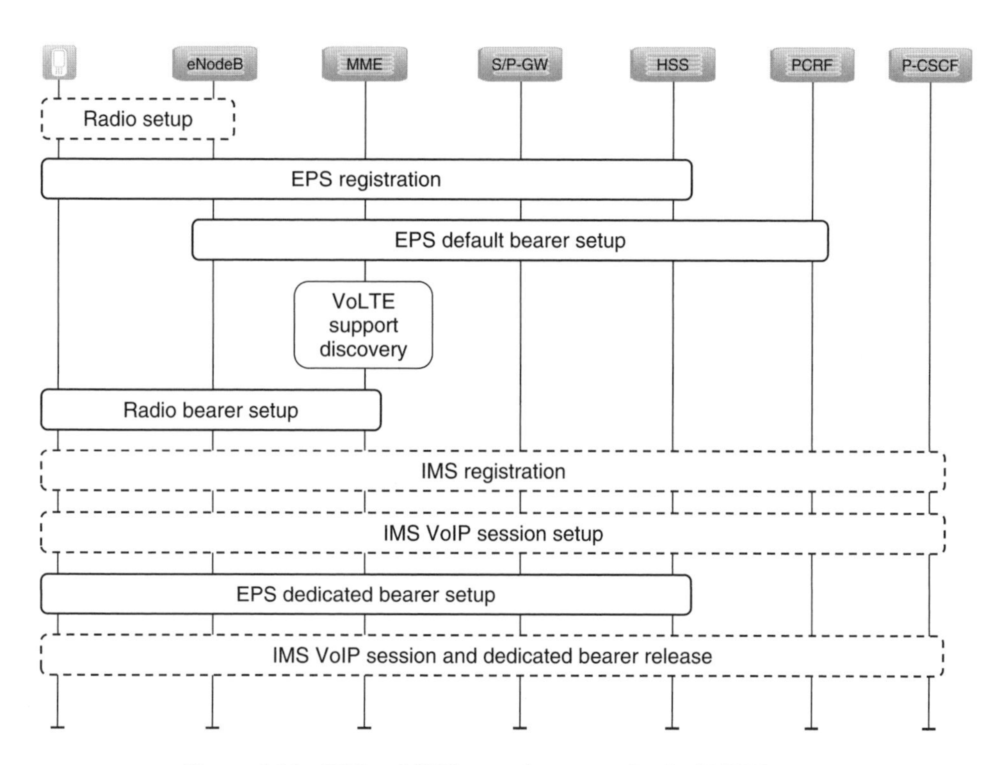

Figure 4.14 EPS and IMS procedure steps for the VoLTE user.

key stored on the USIM and in the authentication centre (AuC). After the authentication the network creates security context between UE and network. Note, the USIM application on a UICC is used for accessing an evolved UMTS terrestrial radio access network (E-UTRAN); subscriber identity module (SIM) application on a UICC is not supported.

For the EPS there are two levels of security association between the UE and the network, one between the UE and eNodeB and another between the UE and MME (Figure 4.15). The NAS security association provides integrity, protection and encryption of NAS signalling using the key hierarchy presented below. The home subscriber server (HSS) provides the security key Kasme from which the MME derives the security keys for NAS ciphering and integrity protection. For the radio interface there is integrity protection for signalling and ciphering for the signalling and the user plane.

This provides a secure IP connection for VoLTE traffic. After security associations are created, the MME updates the subscriber's location to HSS, which responds with stored subscription data. The IMSI is the prime key to the subscription profile stored in the HSS. The EPS subscription profile contains several parameters like IMSI, MSISDN, MME identity and at least one PDN subscription context.

For a VoLTE subscriber the subscription data contains a specific PDN subscription context, which defines default values for the IMS signalling bearer that carries SIP traffic. Table 4.2 gives an example of PDN subscription context for the VoLTE user.

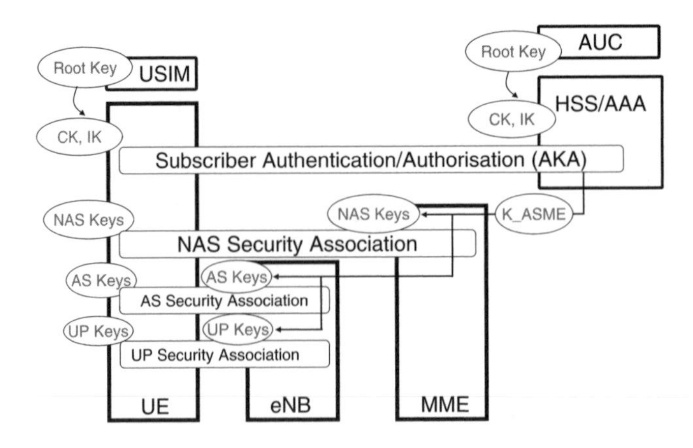

Figure 4.15 EPS security key hierarchy and security associations.

4.3.2.2 EPS Default Bearer Setup

For each PDN connection at least one bearer is created, called the default bearer. For a VoLTE user the default bearer carries all control plane (signalling) traffic coming to/from the IMS APN. When a default bearer is created the P-GW allocates IP address(es) for the PDN connection. The default bearer parameters come from the HSS PDN subscription context presented in Table 4.2. Figure 4.16 shows an example of default bearer setup with related IP address and QoS information.

In the figure 4.6, the P-GW makes the IP address allocation for both Internet protocol version (IPv)4 and IPv6 address types. The IPv6 capable UE must request the IPv4v6 PDN type but the network may provide IPv4, IPv6 or both IPv4 and IPv6 addresses. In the default bearer creation the network provides a full IPv4 address and a globally unique /64 IPv6 prefix, and the UE constructs a full IPv6 address via an IPv6 stateless address autoconfiguration.

The P-GW also sends a proxy call session control function (P-CSCF) address in the PDN connectivity response that is signalled to the UE. The UE shall indicate the request for a P-CSCF address to the network within the protocol configuration options (PCO) information element of the PDN connectivity request message. The network may provide

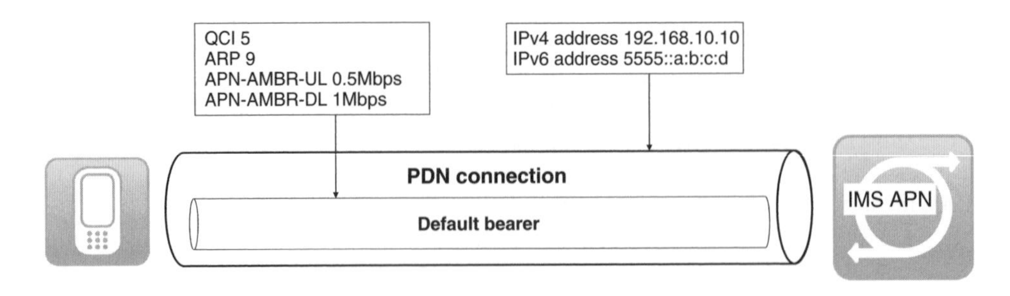

Figure 4.16 Example of default bearer configuration.

Table 4.2 PDN subscription context

Parameter	Purpose	Example value
Context identifier	Index of the PDN subscription context	5
PDN type	Indicates the subscribed PDN Type (IPv4, IPv6, IPv4v6), that is whether IPv4, IPv6 or both addresses are allocated to UE	2 (IPv4v6)
Service-Selection (Access point name)	A label according to DNS naming conventions describing the access point to the packet data network, that is APN network identifier	IMS
EPS subscribed QoS profile	The bearer level QoS parameter values for that APNs default bearer including QCI and ARP (priority and pre-emption)	5 (QCI 5)9 (priority-level)1 (pre-emption_capability_ disabled)0 (pre-emption_ vulnerability_enabled)
Subscribed APN-AMBR	The maximum aggregated uplink and downlink MBR to be shared across all Non-GBR bearers, which are established for this APN. Maximum bandwidth is in bits per second and it contains all the overhead coming from the IP-layer and the layers above, for example IP, UDP, RTP and RTP payload	500 000 (max-requested-bandwidth-UL)1 000 000 (max-requested-bandwidth-DL)
VPLMN address allowed	Specifies whether for this APN the UE is allowed to use the PDN GW in the visited operator network; for VoLTE roaming this must be enabled because P-GW is selected always from the visited PLMN	1 (allowed)

APN-AMBR – APN-aggregate maximum bit rate, MBR – maximum bit rate, DL – downlink, VPLMN – visited PLMN.

a list of P-CSCF IPv4 or IPv6 addresses and the UE must prefer to the IPv6 address type when it discovers the P-CSCF.

The P-GW makes also bearer authorisation during the PDN connection establishment. Figure 4.17 describes signalling where P-GW gets PDN connectivity request with default EPS bearer parameters. The PDN connection establishment is done to the *IMS* APN, which means that the PDN GW is responsible to establish diameter session towards the PCRF over the Gx interface. The request to the PCRF contains at least the user identification, UEs Ipv4 address and/or Ipv6 prefix, Internet protocol-connectivity access network (IP-CAN) type and radio access type. The request may include also other information like the UE time zone and PDN information. The PCRF may ask subscriber specific information from the subscription profile repository (SPR), or the PCRF may use the same rule set for all subscribers, for example on an APN basis. During the PDN connection establishment the PCRF may provide policy and charging control (PCC) rules, bearer

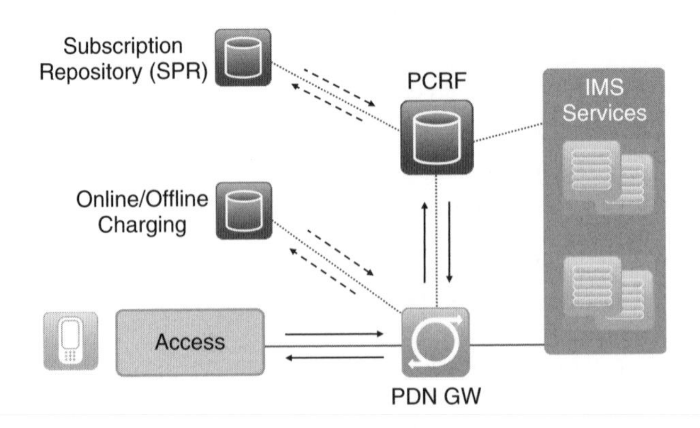

Figure 4.17 PDN connection policy and charging control authorisation.

authorisation and QoS parameters for the PDN connection (IP-CAN session) including a non-GBR, QCI and allocation and retention priority (ARP). More important is that PCRF knows the user's identification and there exists a diameter session, which can be used for dynamic PCC during voice call establishment.

At the same time PDN GW may also connect to a charging system to make proper charging for the user's PDN connection. Because VoLTE signalling and media goes over the IMS APN it is enough to have APN-based charging with a related charging ID.

As part of the default bearer establishment the MME requests also radio access bearer (RAB) establishment. The MME sends bearer parameters to eNodeB to create radio bearer between UE and eNodeB, as shown in Section 4.2.1.

4.3.2.3 IMS Based VoIP Support Discovery

When the VoLTE UE makes an initial attachment it indicates its voice domain preference (preferring IMS PS voice over CS voice or only PS voice capable), usage settings (voice centric or data centric) and SR-VCC capability.

The MME uses the UE provided information, local policy,[1] home network capability, subscription information and the SR-VCC capability of the network to decide whether IMS-based VoIP service can be provided. This decision is signalled back to the UE in the attach accept message using a dedicated parameter (IMS voice over PS session). When the MME has set this indication value equal to true then the UE will realise that the network supports IMS based VoIP and it will initiate IMS registration in order to initiate and receive VoIP communication as described in Section 5.4. If a MME signals that the IMS based VoIP is not available, then the VoLTE UE (as it is voice centric) will attempt CSFB attachment or it will move to UMTS terrestrial radio access network (UTRAN) or GERAN in order to obtain access to voice service.

Although 3GPP specifications do not specify conditions on the network side, which leads to setting the IMS voice over PS session to true or false, the typical decision logic in the MME to set IMS voice over PS session to true is expected to be as follows:

[1] The MME may check, for example based on mobile country code and mobile network code, whether local MME policy would allow this home public land mobile network's (HPLMN's) roamer to have IMS VoLTE.

- UE provided information:
  - UE's usage setting = voice centric;
  - Voice domain preference for E-UTRAN = IMS PS voice preferred, CS voice as secondary;
  - SR-VCC supported (SR-VCC to GERAN/UTRAN capability = SR-VCC from UTRAN HSPA or E-UTRAN to GERAN/UTRAN supported).
- Information available in the network:
  - Existing tracking area (TA) supports IMS VoIP;
  - Network supports SR-VCC and subscription information contains SR-VCC related routing number.

The IMS VoIP indication can be set per TAI or per list of TAIs, and the indication value may change when the UE moves to a new TA.

4.3.3 EPS Dedicated Bearer Setup

The dedicated EPS bearer is needed for voice media because default bearer QoS is not enough. 3GPP has defined the policy and charging control architecture, which is a link between EPC and IMS domains. The policy and charging control is mandatory for IMS voice sessions because the bearer level QoS is based on the IMS voice media session information that is negotiated for each voice call dynamically.

For the voice call the PCRF gets voice session information from the P-CSCF and, based on that information, the PCRF makes QoS provisioning towards P-GW. The PCRF also informs P-CSCF of a bearer level event, for example if QoS cannot be guaranteed or if there is loss of PDN connectivity. The P-GW decides, based on the QCI and ARP pair, that a new dedicated bearer has to be established to provide the required QoS for the voice media. The P-GW initiates dedicated bearer signalling towards access and informs the PCRF after successful establishment.

The dedicated bearer shares the very same IP address(es) as the default bearer but it has a different QoS. For the IMS voice media the dedicated bearer is always GBR, which means that voice media can be guaranteed, or if not then the IMS voice call must be released.

The traffic differentiation is done based on the flow information and P-GW generates a traffic flow template to mobile that the mobile can do a proper bearer selection for uplink traffic.

Figure 4.18 depicts an example of one default bearer with default QoS and one dedicated bearer with QoS applicable for IMS voice media.

4.4 IMS Identification

This section introduces ISIM and different identifiers that are used to identify a user (public user identity), the user's subscription (private user identity), the user's device and a public user identity combination [globally routable user agent URI (GRUU)], the service (public service identity) and IMS network entities. In addition the relationship between different user identities is explained.

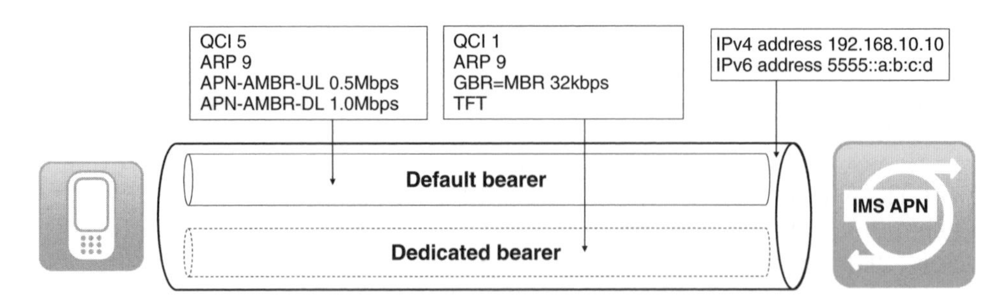

Figure 4.18 Example of default and dedicated bearer configuration.

4.4.1 IP Multimedia Services Identity Module

An ISIM is an application residing on the UICC, which is a physically secure device that can be inserted into and removed from the UE. There may be one or more applications in the UICC, for example USIM and ISIM. The ISIM itself stores IMS-specific subscriber data mainly provisioned by an IMS communication service provider. This data is mainly used when a user registers a device to the IMS. The following data can be stored in ISIM, for example when a user obtains an IMS subscription from a communication service provider:

- Private user identity of the user – this is used in a registration request to identify the user's subscription (see Section 4.4.3 for further information);
- One or more public user identities of the user – this is used in a registration request to identify an identity to be registered and is used to request communication with other users (see Section 4.4.2 for further information);
- Security parameters related to authentication to IMS (such as shared secret and sequence number) and generic bootstrapping architecture (required for Ut authentication, for example for reading/modifying supplementary service settings);
- SMS parameters required for IMS-based messaging;
- List of IMS application reference identifiers which can be used as indicated in IMS registration;
- Address of P-CSCF – this can be used when the access technology does not support dynamic P-CSCF discovery capabilities (not applicable for LTE);
- The name of the entry point of the home network (home network domain name) – this is used in a registration request to route the request to the user's home network;
- Administrative data includes various data which could be used, say, by IMS subscribers for IMS operations or by manufacturers to execute proprietary auto-tests;
- Access rule reference – this is used to store information about which personal identification number needs to be verified in order to get access to the application.

Originally it was assumed that IMS capable devices must be equipped with ISIM but this requirement has been relaxed and currently mobile communication service providers are dominantly allowing access to the IMS with devices equipped with SIM or USIM cards. As EPC access requires the use of USIM, the VoLTE requirement is therefore that the UICC inside the VoLTE UE is equipped either with USIM or with USIM + ISIM.

VoLTE operation is possible just without ISIM as 3GPP has defined how the IMS UE can generate IMS identifiers and home domain name out of data stored in USIM (see Section 4.4.8).

4.4.2 Public User Identity

User identities in IMS networks are called public user identities. They are the identities used for requesting communication with other users. Public identities can be published (e.g. in phone books, Web pages, business cards). IMS users will be able to initiate sessions and receive sessions from many different networks, such as GSM networks and the Internet. To be reachable from the CS side, the public user identity must conform to telecom numbering (e.g. +358501234567). In similar manner, requesting communication with Internet clients, the public user identity must conform to Internet naming (e.g. joe.doe@example.com). The tel universal resource locator (URL) scheme is used to express traditional E.164 numbers in URL syntax. The tel URL is described in (RFC3966), and the SIP uniform resource identifier (URI) is described in (RFC3261) and (RFC3986). User needs to register public user identity (see Section 5.4) before the identity can be used to originating and terminating IMS communication. It is possible to register multiple public user identities through one single UE and even multiple identities through single UE request (see Section 5.4.8). Examples of public user identities are given below.

Example of SIP URI	sip:joe.doe@ims.example.com
Example of tel URL	tel:+358501234567

4.4.3 Private User Identity

The private user identity is a unique global identity defined by the home network communication service provider, which may be used within the home network to uniquely identify the user from a network perspective (3GPP TS 23.228). It does not identify the user herself; on the contrary, it identifies the user's subscription. The private user identity is contained in all registration requests passed from the UE to the home network (see Sections 5.4.1 and 5.4.8) and is authenticated during registration of the user. The private user identity is not used for the routing of SIP messages and it is not present at all in IMS sessions (INVITE) and IMS session-unrelated procedures (e.g. MESSAGE, SUBSCRIBE, NOTIFY). In addition to authentication purposes the private user identities can be used for accounting and administration purposes as well. The private user identity takes the form of a network access identifier (NAI) defined in (RFC2486). An example of private user identity is given below.

Example of NAI	private_user1@home1.csp.net

4.4.4 Relationship between Private and Public User Identities

Here is a basic example to show how different identities are linked to each other. In this example Joe is working for a car sales company and is using a single device for

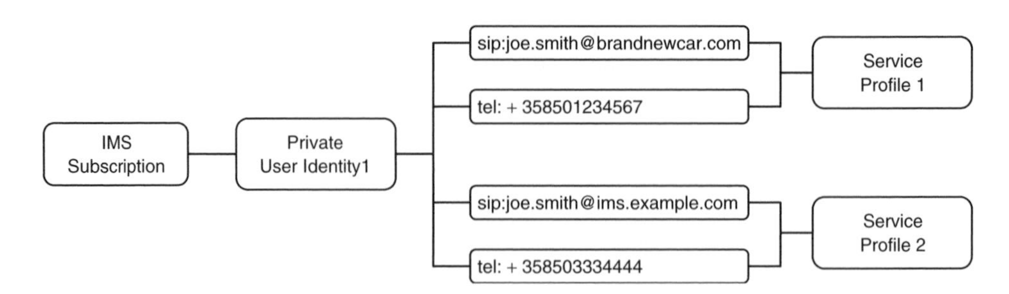

Figure 4.19 Relationship of user identities.

both his work life and his personal life. To handle work-related matters he has two public user identities: sip:joe.smith@brandnewcar.com and tel: +358501234567. When he is off duty he uses two additional public user identities to manage his personal life: sip:joe.smith@ims.example.com and tel: +358503334444. By having two sets of public user identities he can have a totally different treatment for incoming sessions: for example he is able to direct all incoming work-related sessions to a messaging system after 5 p.m. and during weekends and holidays. Joe's user and service-related data are maintained in two different service profiles (see also Figure 4.19). One service profile contains information about his work life identities and is downloaded to the serving call session control function (S-CSCF) from the HSS when needed: that is when Joe registers a work life public user identity or when the S-CSCF needs to execute unregistered services for a work life public user identity. Similarly, another service profile contains information about his personal life identities and is downloaded to the S-CSCF from the HSS when needed. Figure 4.19 shows how Joe's private user identity, public user identities and service profiles are linked together.

IMS also supports more advanced use cases where Joe could have more than one device. So in addition to the identities listed in the previous example Joe could have one additional identity sip:joe@always.net that is shared to all his devices. It means that when someone uses this shared identity to initiate communication Joe can accept the incoming communication with any of his registered devices. According to IMS architecture a shared public used identity must be shared with all private user identities within the IMS subscription (3GPP TS 23.228). Figure 4.20 depicts this configuration (sip:joe@always.net could also point to its own service profile if desired; here it is linked to service profile number 2).

4.4.5 Identification of User's Device

In the IMS public user identities are used to reach the recipient and single public user identity can be shared among a number of devices under a single subscription. This means that it is not possible to identify a specific device when more than one device has been registered with the same public user identity. To reach a particular device a specific identifier called a globally routable user agent URI (GRUU) must be used. For example user Joe has a shared public user identity and if his presence status indicates that he is

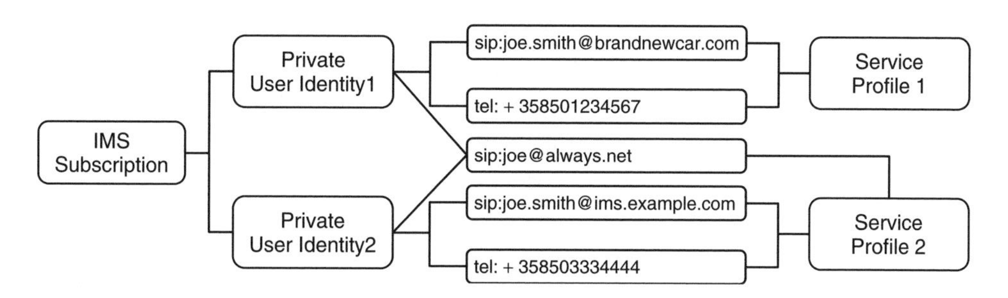

Figure 4.20 Relationship of user identities including shared identity.

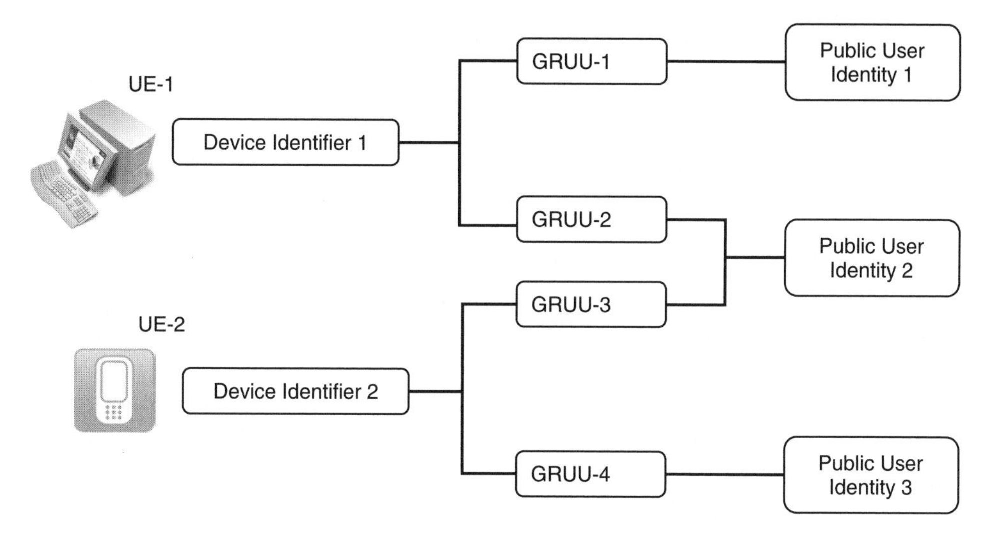

Figure 4.21 Relationship between UE, GRUU and public user identities.

willing to play games with UE1 and that he is willing to accept a video session with UE2 then the GRUU of UE1 (GRUU2 in Figure 4.21) can be used to establish a game session with Joe. Another typical use case for GRUU usage is session transfer from one device to another particular device. Figure 4.21 shows the relationship between UE, GRUU and public user identity.

Two types of GRUU are defined: temporary GRUU and public GRUU. Public GRUU in IMS is a combination of user's public user identity and identifier of the device from which the public user identity is registered to the IMS. The purpose of public GRUU is to enable long lived capability to reach a particular device as it remains the same as long as public user identity–device pair exists. In contrast, temporary GRUU is an identifier that has a limited life time (a new temporary GRUU is created in each IMS registration request) and it keeps the user's public user identity hidden, that is it does not contain the user's public user identity at all. In the example below sip:joe@always.net is the user's public user identity; 'gr' is URI parameter indicating that this SIP URI is actually GRUU, 'urn:uuid' is a uniform resource name (URN) that uniquely identifies this specific device where f81d4fae-7dec-11d0-a765-00a0c91e6bf6 is the unique value identifying particular device.

Example of public GRUU	sip:joe@always.net; gr = urn:uuid:f81d4fae-7dec-11d0-a765-00a0c91e6bf6
Example of temporary GRUU	sip:tgruu.7hs == jd7vnzga5w7fajsc7-ajd6fabz0f8g5@example.com;gr

4.4.6 Identification of Network Entities

In addition to users, network nodes that handle SIP routing need to be identifiable using a valid SIP URI. These SIP URIs would be used when identifying these nodes in the header fields of SIP messages. However, this does not require that these URIs be globally published in a domain name system (DNS), see (3GPP TS 23.228). A communication service provider could name its S-CSCF as follows:

Example of network entity naming	sip:finland.scscf1@ims.example.com

4.4.7 Identification of Services (Public Service Identities)

With the introduction of standardised presence, messaging, conferencing and group service capabilities it became evident that there must be identities to identify services and groups that are hosted by application servers (ASs). Identities for these purposes are also created on the fly: that is they may be created by the user on an as-needed basis in the AS and are not registered prior to usage. Ordinary public user identities were simply not good enough; so, Release 6 introduced a new type of identity, the public service identity. Public service identities either take the form of a SIP URI or are in tel URL format. Section 4.6.3.6 shows an example use case where user creates an ad-hoc conference using a public service identity (conference factory URI) and an AS further shares the dynamic public service identity as a conference URI.

4.4.8 Identification Without ISIM

Sections 4.4.2 and 4.4.3 explained the concepts of public user identity and private user identity. It was assumed that these identities are stored in an ISIM application. When the IMS is deployed there is a number of UE in the market place that are not equipped with the ISIM application; therefore, a mechanism to access the IMS without the ISIM was developed. In this model, private user identity, public user identity and home domain name are derived from an IMSI. This mechanism is suitable for UE that has a USIM application.

4.4.8.1 Derived Private User Identity

The private user identity derived from the IMSI is built according to the following steps (3GPP TS 23.003):

1. Use the whole string of digits as the username part of the private user identity.
2. Form the domain part of private user identity by using predefined syntax, ims.mnc.mcc.3gppnetwork.org, where MNC and MCC are extracted from IMSI (first three leading digits identifies MCC and next 2 or 3 digits identifies MNC).[2]

For example:

```
IMSI in use: 244911234567; where:
MCC: 244;
MNC: 91;
MSIN: 1234567; and
Private user identity is:
244911234567@ims.mnc091.mcc244.3gppnetwork.org
```

4.4.8.2 Temporary Public User Identity

If there is no ISIM application to host the public user identity, a temporary public user identity will be derived, based on the IMSI. The temporary public user identity will take the form of a SIP URI, 'sip:user@domain'. The user and domain part are derived similarly from the method used for private user identity (3GPP TS 23.003). Following our earlier example a corresponding temporary public user identity would be:

```
sip:244911234567@ims.mnc091.mcc244.3gppnetwork.org
```

When a temporary public user identity is used in an IMS initial registration it is placed in To and From headers (see Section 5.4.1). A valid IMS public user identity will be implicitly registered and the S-CSCF will deliver it to the UE (in P-Associated-URI header) when it accepts the registration (see Section 5.4.8). After the initial registration, the UE will only use the implicitly registered public user identity(s). It is strongly recommended that the temporary public user identity is set to 'barred' for IMS non-registration procedures so that it cannot be used for IMS communication.

4.4.8.3 Home Network Domain Name

A home network domain name is derived in the same way as the domain part of temporary private user identity. So in our example it would be:

```
ims.mnc091.mcc244.3gppnetwork.org
```

4.5 IMS Service Provisioning

Strictly speaking the IMS is not a service in itself; on the contrary, it is a SIP-based architecture for enabling an advanced IP service and application on top of the PS network. IMS provides the necessary means for invoking services; this functionality is called here as 'service provisioning'. Figure 4.22 gives an overview of service provisioning.

[2] If the MNC is two digits then a zero shall be added at the beginning to make it three digits long.

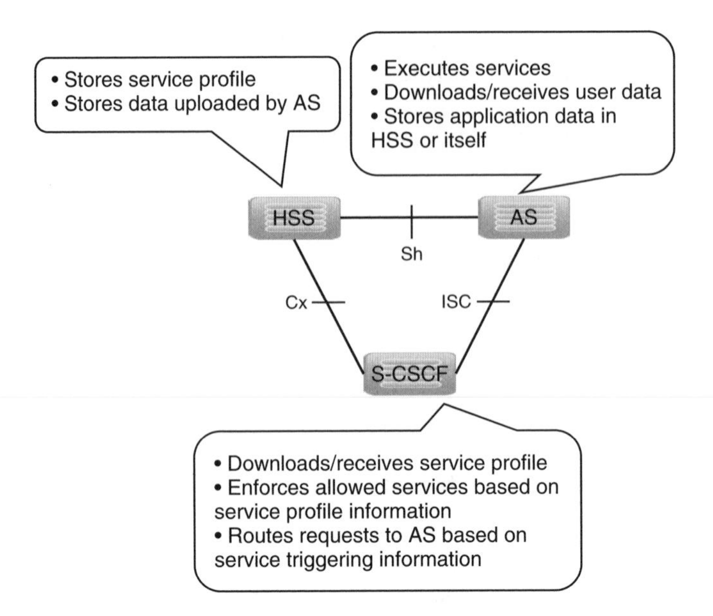

Figure 4.22 IMS service provisioning overview.

The HSS as a master database contains permanently user's IMS subscription information and one part of this information is IMS service profile (see also Section 5.2 and especially Figure 5.2). In addition may act as a permanent storage for application specific data uploaded by AS. This data may also contain such data which is not understood semantically by the HSS.

4.5.1 Enforcement of Allowed Services

While serving a user, the S-CSCF uses the downloaded service profile to enforce allowed services based on the configured media policy and IMS communication service identifier policy.

- Media policy information contains an integer that identifies a subscribed media profile in the S-CSCF (e.g. allowed session description protocol (SDP) parameters). This information allows communication service providers to define different subscriber profiles in their IMS networks. They may define different customer classes, such as gold, silver and bronze. Gold could mean that a user is able to make video calls and all ordinary calls. Silver could mean that a user is able to use wideband adaptive multi-rate (AMR) as a speech codec, but they are not allowed to make video calls and so on. Transferring just the integer value between the HSS and the S-CSCF saves the storage space in the HSS and optimises the usage of the Cx reference point.
- IMS communication service identifier policy contains a list of service identifiers that identifies which IMS communication service user is entitled to use, for example IMS MMTel. Based on the provided list the S-CSCF enforces usage of identifiers in SIP signalling.

4.5.2 *Service-Triggering Information*

Service-triggering information is presented in the form of initial filter criteria (see Figure 4.23). The initial filter criteria describe when an incoming SIP message is further routed to a specific AS. The service profile may contain both user-specific service-triggering information which is coded as initial filter criteria and a reference value to initial filter criteria which are locally administrated and stored in S-CSCF. The latter is called shared initial filter criteria and it is encoded as an integer value where the integer value only has a meaning inside a single communication service provider's network. For example, value 1 could point triggers that take care of routing requests to open mobile alliance (OMA) instant messaging (IM), push to talk over cellular (PoC), Presence and XML document management (XDM) applications, and value 2 could point triggers that take care of routing requests to IP centrex.

Whenever a user obtains a VoLTE subscription then it is fair to assume that her subscription will contain a number of voice supplementary services which are provided by a telephony application server (TAS) as described in Section 4.6.3. Therefore, a

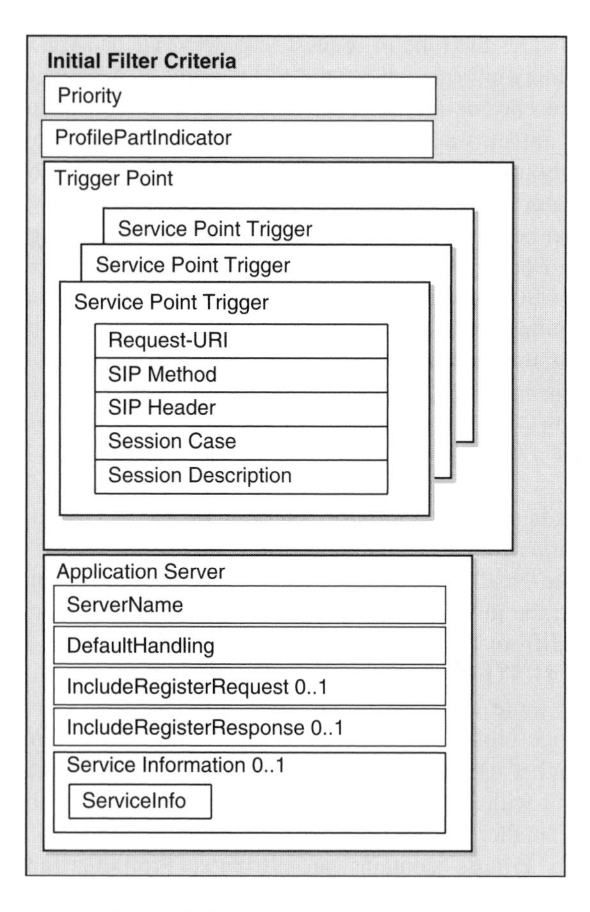

Figure 4.23 IMS initial filter criteria.

communication service provider needs to include initial filter criteria to her service profile which will ensure that S-CSCF routes both incoming and outgoing voice session to the TAS for service execution.

When constructing the initial filter criteria the communication service provider needs to consider these questions:

- What is the priority of an initial filter criterion?
- What is a trigger point?
- Is this initial filter criteria for a registered user or an unregistered user?
- What is the correct AS when the trigger point is met?
- What should be done if the AS is not responding?

The trigger point is used to decide whether an AS is contacted. It contains one to multiple instances of a service point trigger, as shown in Figure 4.23. The service point trigger comprises these items:

- Request-URI – identifies a resource that the request is addressed to (e.g. sport-news@ims.example.com).
- SIP method – indicates the type of request (e.g. INVITE or MESSAGE).
- SIP header – contains information related to the request. A service point trigger could be based on the presence or absence of any SIP header or the content of any SIP header. The value of the content is a string that is interpreted as a regular expression. A regular expression could be as simple as a proper noun (e.g. John) in the FROM header that indicates the initiator of the request.
- Session case – can be any one of four possible values, originating, terminating, originating unregistered or terminating unregistered, that indicate whether the filter should be used by the S-CSCF that is handling the originating service, terminating service or originating/terminating for an unregistered end user service. An originating case refers to when the S-CSCF is serving the calling user. A terminating case refers to when the S-CSCF is serving the called user.
- Session description – defines a service point trigger for the content of any SDP field within the body of a SIP method. Regular expressions can be used to match the trigger.

If the trigger point is met then the initial filter criteria contain instructions on what to do next. These instructions contain the address of the AS, what to do if contact with the AS fails (either terminate or allow session to proceed) and may contain further instructions to the S-CSCF when the initial filter criterion is set for IMS registration (i.e. pass service information received from HSS to AS, deliver REGISTER request from UE to AS and pass 200 OK (of REGISTER) request to the AS).

A user may have more than one initial filter criteria in her service profile therefore communication service profile needs to assign an unique priority number for each initial filter criteria. The higher the priority number the lower the priority of the filter criteria is; that is a filter criteria with a higher value of priority number shall be assessed after the filter criteria with a smaller priority number have been assessed.

Each initial filter criteria contains an attribute, ProfilePartIndicator, which tells whether particular initial filter criteria is to applied when the user is either registered or unregistered.

4.5.3 Selection of AS

The S-CSCF assesses the filter criteria for the initial request alone, according to the following steps (3GPP TS 24.229):

1. Check whether the public user identity is barred; if not, then proceed.
2. Check whether this request is an originating request or a terminating request.
3. Select the initial filter criteria for a session case (originating, terminating or originating/terminating for an unregistered end user).
4. Check whether this request matches the initial filter criterion that has the highest priority for that user by comparing the service profile with the public user identity that was used to place this request:
   a. If this request matches the initial filter criterion, then the S-CSCF will forward this request to that AS, check to see whether it matches the next following filter criterion of lower priority and apply the filter criteria on the SIP method received from the previously contacted AS.
   b. If this request does not match the highest priority initial filter criterion, then check to see whether it matches the following filter criterion's priorities until one does match.
   c. If no more (or none) of the initial filter criteria apply, then the S-CSCF will forward this request based on the route decision.

There exists one clear difference in how the S-CSCF handles the originating and terminating initial filter criteria. When the S-CSCF realises that an AS has changed the Request-URI in the case of terminating initial filter criteria it has three options:

- Route the request based on the changed value of the Request-URI;
- Use and execute the initial filter criteria for the UE-originating case after retargeting;
- Continue to use the initial filter criteria for the UE terminating case.

In an originating case the S-CSCF will continue to evaluate initial filter criteria until all of them have been evaluated.

If the contacted AS does not respond, then the S-CSCF follows the default-handling procedure associated with initial filter criteria: that is either terminate the session or let the session continue, based on the information in the filter criteria. If the initial filter criteria do not contain instructions to the S-CSCF regarding the failure to contact the AS, then the S-CSCF will let the call continue as the default behaviour (3GPP TS 24.229).

4.5.4 AS Behaviour

4.5.4.1 Service Execution

Section 4.5.3 described how the request is routed to an AS. After receiving the request the AS initiates the actual service. To carry the service out the AS may act in four different modes:

- Terminating user agent (UA) – the AS acts as the UE. This mode could be used, for example to provide a voice mail service.

- Redirect server – the AS informs the originator about the user's new location or about alternative services that might be able to satisfy the session.
- SIP proxy – the AS processes the request and then proxies the request back to the S-CSCF. While processing, the AS may add, remove or modify the header contents contained in the SIP request according to the proxy rules specified in (RFC3261).
- Third-party call control/back to back UA – the AS generates a new SIP request for a different SIP dialogue, which it sends to the S-CSCF. This could be, for example TAS providing an ad-hoc multi party conference (see Section 4.6.3.6).

In addition to these modes, an AS can act as an originating UA. When the application is acting as an originating UA it is able to send requests to the users: for example a conferencing server may send SIP INVITE requests to a pre-defined number of people at 9 a.m. for setting up a conference call. Another example could be a news server sending a SIP MESSAGE to a soccer fan to let him know that his favourite team has scored a goal.

4.5.4.2 User and Service Specific Data

There are different sources where the AS can receive user- and service-specific data. Users can utilise the Ut reference point (see Figure 3.18) to store application specific data in the AS, for example supplementary service data (see Section 4.6.3.11). A communication service provider can use operation and management tools to configure information to the AS. The AS can use the Sh reference point (see Figure 3.18) to store data in the HSS to support its service logic. Via the Sh reference point the AS can also download and receive all kinds of data available in the HSS or available via the HSS. This information contains, for example IMS registration status, CS/PS network [MSC, MME, serving GPRS support node (SGSN)] status, public user identities, initial filter criteria, the S-CSCF name serving the user, addresses of the charging functions, UE IP address information, the UE's SR-VCC capability information, CS domain routing number, information which radio access type (e.g. E-UTRAN) is serving the UE and whether or not IMS voice over the PS session is supported at the current geographic area where the UE is currently located, session transfer number for SR-VCC, Internet protocol short message gateway (IP-SM-GW) number and even location information from the CS and PS domains (3GPP TS 29.329).

4.5.5 Service Provisioning in Action

Figure 4.24 summarises the service provisioning in action. In step 1 either the originating or terminating request arrives at the S-CSCF. If the S-CSCF is serving an unregistered user and the service profile(s) is(are) not already available then the S-CSCF downloads the profile in step 2. In step 3 the S-CSCF executes enforcement of the allowed services as described in Section 4.5.1 and evaluates the initial filter criteria. As an outcome of matching initial filter criterion the request is forwarded to the AS for service execution (step 4). The AS may retrieve additional information from the HSS (step 5) as described in Section 4.5.4.2 and the AS executes the service in step 6. In step 7 the outcome of execution is communicated back to the S-CSCF. Depending on result of service execution the

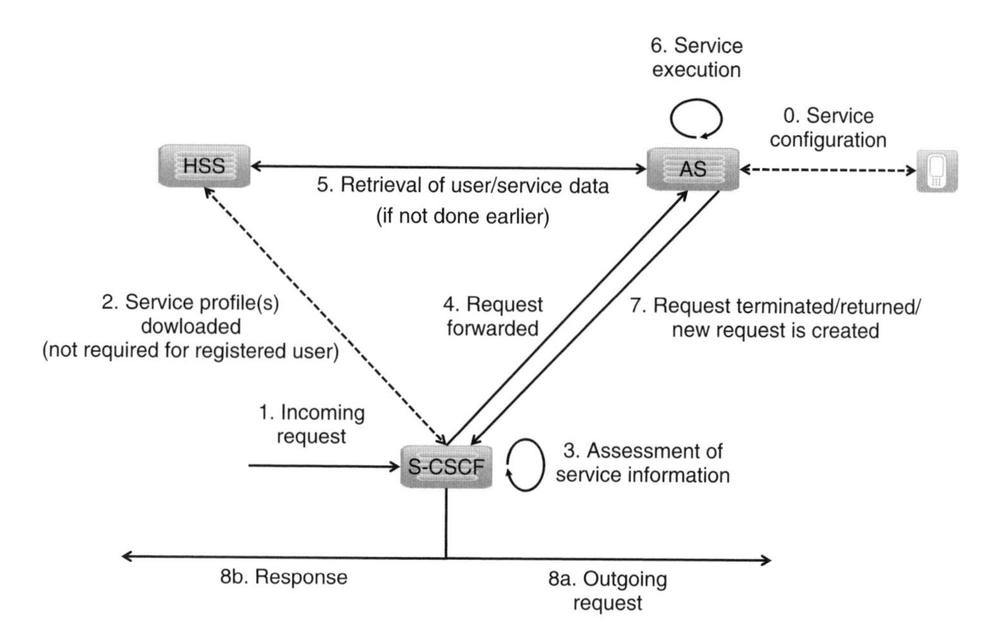

Figure 4.24 IMS service execution overview.

S-CSCF executes the next steps (continue initial filter criteria evaluation, route request forward or pass response back to requestor).

For a VoLTE subscriber, a communication service provider needs to create at least some initial filter criteria to make sure that all voice sessions gets routed to TAS and SMS over IP messages are correctly delivered to the IP-SM-GW. Table 4.3 shows the basic required initial filter criteria settings for VoLTE user. The first two criteria are required to cover

Table 4.3 VoLTE initial filter criteria examples

Element of filter criteria	Filter criterion 1	Filter criterion 2	Filter criterion 3	Filter criterion 4
SPT: SIP method	Invite	Invite	Invite	Message
SPT: Session case	Terminating	Terminating	Originating	Originating
Further SPT	Session description: m = audio	Session description: m = audio	Session description: m = audio	SIP header: content-type:application/vnd.3gpp.sms Request-URI: +358508888888
Application server	sip:tas1@example.com	sip:tas1@example.com	sip:tas1@example.com	sip:ipsmgw1@example.com
Priority	1	2	3	4
Profile part indicator	Registered	Unregistered	Registered	Registered

SPT – Service point trigger.

mobile terminating voice sessions. The third criterion is used to make sure that VoLTE user initiated voice sessions will receive originating voice services. The fourth criterion is set to ensure that SMS over IP messages are interworking at the originating network and not passed towards remote networks which may not have messaging interworking technology in place.

To complete IMS service provision let us verify that the above filter criteria 1 and 3 work for our example VoLTE user, Kristiina (see Section 5.1).

4.5.5.1 Terminating VoLTE Session

When Kristiina receives a request to engage on voice communication her S-CSCF will receive a SIP INVITE targeted to her. The INVITE will contain among other things following key pieces of information:

```
INVITE sip:kristiina@example.com SIP/2.0
P-Asserted-Identity:sip:alice@example.com
Route:<sip:scscf1.example.com;lr>
Content-Type: application/sdp
m=audio 3420 RTP/AVP 97 98
```

The S-CSCF detects that the filter criterion 1 matches, because:

- The SIP method is INVITE.
- The INVITE request is targeted to the terminating user – the S-CSCF knows this based on the content of the Route header. It does not contain the value which was communicated in the Service-Route header during the registration to the UE and P-CSCF (see Section 5.4.8).
- The SDP payload contains a dedicated line stating that this request is for audio communication.
- Kristiina is registered.

4.5.5.2 Originating VoLTE Session

When Kristiina wants to engage in voice communication her UE will send a SIP INVITE to the network as described in detail in Section 5.5. This invite will find its way to the S-CSCF serving Kristiina. Among other things, the INVITE will contain the following key pieces of information:

```
INVITE sip:jennifer@csp.com SIP/2.0
P-Asserted-Identity: sip:kristiina@example.com
Route:<sip:orig@scscf1.home.fi;lr>
Content-Type: application/sdp
m=audio 3458 RTP/AVP 97 98
```

The S-CSCF detects that filter criterion 3 matches, because:

- The SIP method is INVITE.

- The INVITE request is received from the originating user – the S-CSCF knows this from the user part it set in its Service-Route header entry (see Section 5.5.8) and which is now returned in the Route header.
- The SDP payload contains a dedicated line stating that this request is for audio communication.
- Kristiina is registered.

4.6 IMS Multimedia Telephony

4.6.1 Introduction

The MMTel is a blended standardised multimedia service suite in the IMS and provides support for mobile as well as for fixed users. The MMTel service allows users to establish communications between them and enrich that communication by enabling supplementary services.

The IMS MMTel service consists of two principal parts:

- Basic communication (Section 4.6.2);
- Optional supplementary services (Section 4.6.3).

The need for IMS supplementary services was discussed in 3GPP Release 5, and at that time 3GPP decided that there was no need to standardise supplementary services as SIP protocol enables basic capabilities such as communication forwarding, identity services and communication hold. After ETSI telecommunications and internet converged services and protocols for advanced networking (TISPAN) made a decision to use the IMS core network as a platform for their next generation network (NGN) they started to define supplementary services for the IMS. The first set of supplementary services was completed as part of TISPAN Release 1. In parallel 3GPP as part of the work on 3GPP Release 7 defined the so-called IMS MMTel service which has reused some of the TISPAN supplementary services and made them available in general to the IMS. In TISPAN Release 2, TISPAN continued to work on supplementary services for IMS, but after completion of TISPAN Release 2 the responsibility for all specifications relevant for IMS core network including the specifications for supplementary services was transferred from TISPAN to 3GPP as part of the effort on a common IMS (see also Section 3.4.2 and Figure 4.25).

The One Voice initiative defined a profile for the IMS for voice and SMS which contains a minimum set of features to be supported by a cellular UE and an IMS network which has been published as GSM association (GSMA) PRD (IR.92). Part of the scope was to define a subset of MMTel supplementary services and related capabilities so that a telephony service as known from 2G/3G networks can be built. The 3GPP Release 8 set of specifications for MMTel do not provide capabilities for the end user to manipulate the supplementary service settings from both 2G/3G and MMTel UE in a consistent manner, that is it does not provide a capability to simulate the capabilities of the supplementary services as known from 2G/3G. In other words a 1 : 1 synchronisation between 2G/3G supplementary service settings and full blown IMS MMTel is not possible because MMTel contains options which are not applicable in 2G/3G. It was recognised that this can

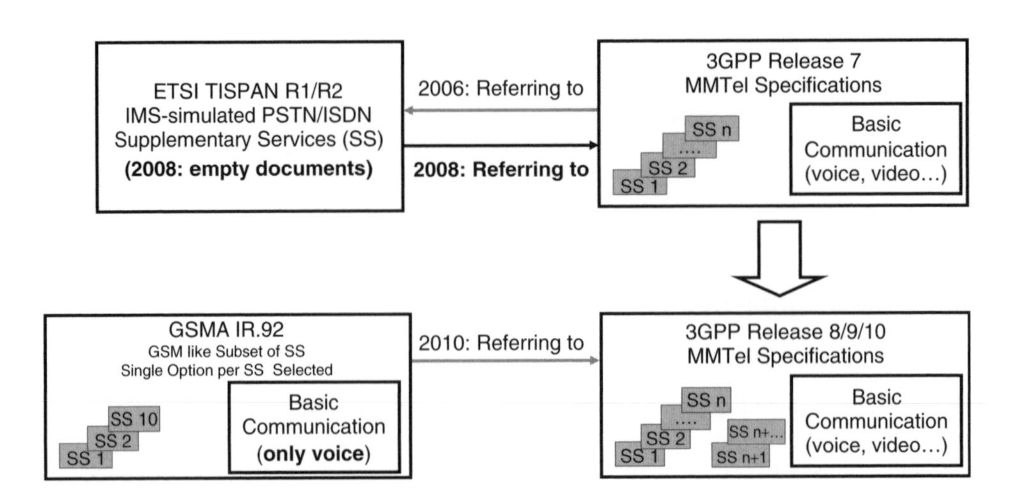

Figure 4.25 Supplementary services in TISPAN, 3GPP MMTel and GSMA (IR.92).

lead to an unwanted end user experience and a solution for this was specified in 3GPP Release 9. In order to provide the possibility to profile the existing MMTel supplementary services and its options, the specifications for the impacted supplementary services and the specification for XCAP over Ut (3GPP TS 24.623), were enhanced to allow the UE to get information about which supplementary services and which actions and conditions for a supplementary service are supported. This means that it is possible to build services such that only those settings which can be mapped to the 2G/3G supplementary service data model are allowed and the UE is informed accordingly. As part of the solution, the network can reject service modification requests which violate configuration limitations in the network. The solution does not by any means limit already existing standardised MMTel capabilities. It rather provides the possibility to define a profile of supplementary services like GSMA (IR.92).

4.6.2 Multimedia Communication

Typically MMTel is a service using speech and speech combined with other media components, but the service is not limited to always include speech, it also caters for other media or combinations of media (e.g. text and video). MMTel service includes the following standardised media capabilities:

- Full duplex speech;
- Real time video (simplex, full duplex), synchronised with speech if present;
- Text communication;
- File transfer;
- Video clip sharing, picture sharing, audio clip sharing; transferred files being displayed/replayed on the receiving terminal for specified file formats;
- User datagram protocol (UDP) transport layer based transfer of fax.

Based on these capabilities different types of communication use cases could be realised, for example voice calls, voice call enriched with video (unidirectional or bidirectional), basic voice call enriched with sharing, basic voice call enriched with file transfer, basic voice call enriched with video and text communication, basic voice call enriched with video and text communication and file transfer, text communication (chat), file transfer message session relay protocol (MSRP), sharing (image, video). 3GPP has defined a dedicated IMS communication service identifier (ICSI), 3gpp-service.ims.icsi.mmtel, which UE adds to its IMS registration (see Section 5.4.1) and IMS session (see Section 5.5.1) to indicate that desired service is MMTel. For speech, video real-time text and fax 3GPP has defined rules for media encoding as summarised in Table 4.4 (3GPP TS 26.114). For file transfer, sharing and chatting MSRP (RFC4975) is required. All codecs in Table 4.4 use the RTP (RFC3550) over UDP and IP as transport. They apply specific payload formats within RTP, for example (RFC4867) for AMR and adaptive multi-rate wideband (AMR-WB). They also apply a so-called RTP profile, which is signalled in the SDP (RFC4566) as transport parameter. RTP/AVP(attribute value pair) (RFC3551) is the oldest such RTP profile developed by IETF and in widespread use. RTP/AVPF (audio-visual profile with feedback) (RFC4585) is an enhanced profile. Applicable codecs for media streams are negotiated using SDP within SIP signalling and applying the offer-answer procedures (see Section 5.4.3).

4.6.3 Supplementary Services

In order to provide supplementary services within the IMS, the IMS core network will include a Telephony Application Server (TAS) in the call chain. The TAS is a SIP application server that is connected via the IMS service control (ISC) interface and

Table 4.4 Codecs used for MMtel media

Media type	3GPP requirements (Release 10)
Speech	AMR narrowband (NB) is mandatory Wideband (16 kHz sampling frequency) is optional. If supported then AMR-WB is mandatory. DTMF support is mandatory.
Video	ITU-T recommendation H.264/MPEG-4 (part 10) advanced video coding (AVC) 24 constrained baseline profile (CBP) level 1.2 is mandatory.
Real-time text	ITU-T recommendation T.140
Fax	ITU-T recommendation T.38
Media transport for speech, video and text	In Rel-7 and Rel-8, support of RTP/AVP mandated for all media types and support of RTP/AVPF is mandated for video and recommended for other media types. In Rel-9 support of both RTP/AVPF and RTP/AVP is mandated for all media.

DTMF – Dual-tone multifrequency.

provides the network support for the supplementary services of MMTel. Depending on whether the supplementary service applies to the originating or terminating side, a TAS will be included on the originating and/or terminating side network.

In addition to the use of SIP for call control of the supplementary services the TAS needs to provide a configuration interface to allow the user to configure its service settings. Via this interface the user can configure the behaviour of some of the supplementary services. IMS has defined the following possibilities:

- Using XCAP via the Ut interface using;
- Using SIP based user configuration.

Further it is possible to employ non standardised means to configure the behaviours of supplementary services, for example a Web-based interface.

Figure 4.26 shows the principle architecture and protocols used for MMTel. AP in Figure 4.26 stands for authentication proxy. It is a 3GPP function that takes care of authenticating the user when the user needs to send XCAP requests and it takes care of finding the right XCAP server (MMTel, IM, Presence, etc.) providing the required service functionality.

Since the start of specifying MMTel in 3GPP Release 7, the list of supported supplementary services in MMTel has grown with every 3GPP Release. Most of the supplementary services are similar to supplementary services specified for CS speech. Table 4.5 lists standardised IMS supplementary services. The last column in Table 4.5 indicates whether a particular supplementary service is included in the GSMA (IR.92) IMS profile for voice and SMS. In the following sections we describe the supplementary services which are included in GSMA (IR.92). GSMA (IR.92) includes additional restrictions for communication diversion (CDIV) and communication barring services. These restrictions are covered in Section 4.6.3.11.

4.6.3.1 Originating Identification Presentation

The public user identity is conveyed within three different headers within a SIP INVITE request:

- The From header, which can be set to any value the calling user wants;
- The P-Preferred-Identity header, in which the calling user can indicate one of its public user identities, that is the identity that it wants to be used for the call;
- The P-Asserted-Identity header, in which the authorised identity of the user is transported.

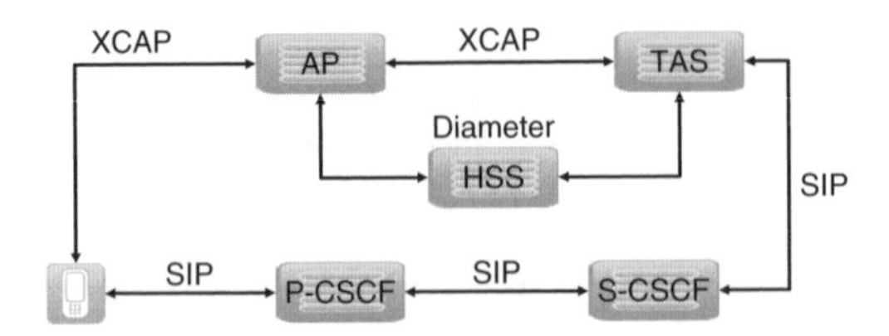

Figure 4.26 Basic IMS architecture in support of multimedia telephony.

Table 4.5 IMS supplementary services

Supplementary service	Description	GSMA (IR.92) functionality
Release 7 baseline		
Originating identification presentation (OIP)	Service provides the terminating user with the identity of the originating user	Yes
Originating identification restriction (OIR)	Service enables the originating user to prevent presentation of its identity information to the terminating user	Yes
Terminating identification presentation (TIP)	Service provides the originating user with the possibility of receiving identity information in order to identify the terminating user	Yes
Terminating identification restriction (TIR)	Service offered to the connected user which enables the connected user to prevent presentation of the terminating identity information to the originating user	Yes
Communication hold (HOLD)	Service enables a user to suspend media within a session, and resume that media at a later time	Yes
CONFerence calling (CONF)	Service enables a user to participate in and control a simultaneous communication involving a number of users	Yes
Communication DIVersion services (CDIV)	Service (aka communication forwarding) allows the user to re-direct an incoming request that fulfils certain provisioned or configured conditions to another destination	Yes
Communication barring (CB)	Service allows users to selectively block session attempts	Yes
Message waiting indication (MWI)	Service enables the application server to indicate to the user, that there is at least one message waiting	Yes
Explicit communication transfer (ECT)	Provides a user involved in a communication to transfer that communication to a third party	No
Release 8 additions		
Communication waiting (CW)	Allows an user that has activated the service and is currently having an active call to be informed about an additional incoming call	Yes
Malicious communication identification (MCID)	Allows a user to identify malicious communication	No
Completion of communications to busy subscriber (CCBS)	Enables an originating user, encountering a busy destination B, to have the communication completed without the user having to manually initiate a new communication attempt when the destination B becomes not busy	No

Table 4.5 (*continued*)

Supplementary service	Description	GSMA (IR.92) functionality
Advice of charge (AOC)	Allows a served user to be informed of IP multimedia session related charging information	No
Closed user group (CUG)	Allows a user to form groups of members, whose communication profile is restricted for incoming and outgoing communications	No
Completion of communications on no reply (CCNR)	Enables an originating user, encountering a destination B which does not answer the communication (no reply), to have the communication completed without the user having to manually initiate a new communication attempt when the destination becomes not busy after having initiated and completed a new communication	No
Customised alerting tone (CAT)	Allows a user to customise the media which is played to the calling user during alerting of the called user	No
Flexible alerting (FA)	Service enables a call to a pilot identity to branch the call into several legs to alert several termination addresses (group members) simultaneously	No
Release 9 addition[3]		
Customised ringing signal (CRS)	Service allows the user to customise the media which is played to herself/himself during the alerting phase of an incoming call	No

When setting up the call, the user indicates the public user identity in the P-Preferred-Identity Header and the INVITE request will look like the following:

```
INVITE sip:jennifer@csp.com SIP/2.0
From: "Kristiina" <sip:kristiina@example.com>;tag=171828
P-Preferred-Identity: <sip:kristiina@example.com>
```

The P-CSCF checks, based on registration state information that it has learnt due to its subscription to the user's registration state event package (see Section 5.4.10), whether the indicated identity is a valid public user identity of the user:

- If the identity is valid, it replaces the P-Preferred-Identity header with the P-Asserted-Identity header, including the same value.
- If the identity is not valid, it removes the P-Preferred-Identity header and puts the value of the default public user identity learned during the IMS registration (see

[3] In addition to new service, CRS, Release 9 introduces service capability discovery mechanism and some service definition changes.

Section 5.4.8–first identity received in the P-Associated-URI header is the default identity) into the P-Asserted-Identity header.

```
INVITE sip:jennifer@csp.com SIP/2.0
From: "Kristiina" <sip:kristiina@example.com>;tag=171828
P-Asserted-Identity: <sip:kristiina@example.com>
```

The P-Asserted-Identity header, is used to convey a trusted identity of the caller, which is not only used for the OIP supplementary service, but also for identification of the authenticated user within the IMS network. The P-Asserted-Identity header will be available in the terminated UE when the SIP signalling has traversed through a trusted signalling network and no communication service provider has disabled OIP from the user. In the case when a CSCF detects that the next hop is untrusted, the CSCF will remove the P-Asserted-Identity header, which prevents the OIP supplementary service.

4.6.3.2 Originating Identification Restriction

Originating identification restriction (OIR) is a service that an originating user can use in order not to expose the user's identity to other users. Users can activate this on a request basis or set default configuration to the TAS within the user's IMS network. In order to restrict the presentation of the identity in the From header, the From header must be set to 'anonymous'. If the calling user wants to apply OIR only for a specific call, the user can set the From header immediately to 'anonymous' when sending out the SIP INVITE request.

```
INVITE jennifer@csp.com SIP/2.0
From: "anonymous" <sip:anonymous@example.com>;tag=xyz
```

It might well be that the user has subscribed to the OIR service in general and wants the service always to be invoked, regardless of what the originating UE indicates in the From header. In this case, the TAS of the calling user will apply the service, by changing the From header from the value set by the user to 'anonymous'. In order to achieve this, the TAS will act as a SIP back to back user agent (B2BUA). The P-Asserted-Identity header nevertheless cannot be set to 'anonymous', as it is essential for the user's identification within the network. Therefore an additional SIP header called 'privacy' is used in order to indicate that the originating user does not want the asserted identity to be shown to the called user. To achieve OIR for a single call, the calling user will include the 'privacy' header set to the value 'id' in the initial invite request, which will force the terminating P-CSCF to remove the P-Asserted-Identity header from the SIP request.

```
INVITE sip:jennifer@csp.com SIP/2.0
P-Asserted-Identity: <sip:kristiina@example.com>
Privacy: id
```

If the calling user has activated the OIR supplementary services on a permanent basis, the TAS will include the 'privacy: id' header within the SIP INVITE request. A user can permanently activate the OIR service by sending a XCAP message to the TAS, indicating that the presentation should be restricted. The user can also deactivate the service, by sending a XCAP message to the TAS, indicating that the presentation is allowed.

4.6.3.3 Terminating Identification Presentation

The called user will be identified by a P-Asserted-Identity header, which is sent in the first SIP response towards the originating user. The P-Asserted-Identity header is set by the P-CSCF of the called user, based on the value received in the P-Preferred-Identity header from the called terminal in the SIP response (see Section 4.6.3.1).

```
SIP/2.0 183 Session Progress
P-Asserted-Identity: <sip:jennifer@csp.com>
```

As long as all IMS entities between the two users are trusting each other, the P-Asserted-Identity header will stay within the 183 (Session Progress) response and will be delivered to the calling user's terminal, where it can be presented to the called user.

4.6.3.4 Terminating Identification Restriction

The terminating identification restriction (TIR) supplementary service enables the called user to restrict the presentation of the terminating public user identity in the same way as the originating user can restrict the presentation of the calling public user identity (as described for the OIR supplementary service – see Section 4.6.3.2). In order to restrict the presentation of the identity, the called user adds a Privacy header with the value 'id' into every SIP response that it sends and that conveys the P-Preferred-Identity header. The P-CSCF of the called user then will include the P-Asserted-Identity header for the called user and will send the Privacy header forward unchanged. The P-CSCF of the calling user will remove the P-Asserted-Identity header, based on the presence of the 'Privacy: id' header and will not forward the terminating identification to the called user.

```
SIP/2.0 Session Progress
P-Asserted-Identity: <sip:jennifer@csp.com>
Privacy: id
```

The terminating user can also subscribe on a permanent basis to the TIR supplementary service, by sending a XCAP message to the called user's TAS, indicating that presentation of the terminating identification is restricted.[4] Based on this, the TAS will add the 'Privacy: id' header in every response from the called user which includes the P-Asserted-Identity header.

4.6.3.5 Communication Hold

The communication hold supplementary service enables a user to suspend the media stream(s) of an established IP multimedia session, and resume the media stream(s) at a

[4] XCAP service configuration is not required in GSMA (IR.92).

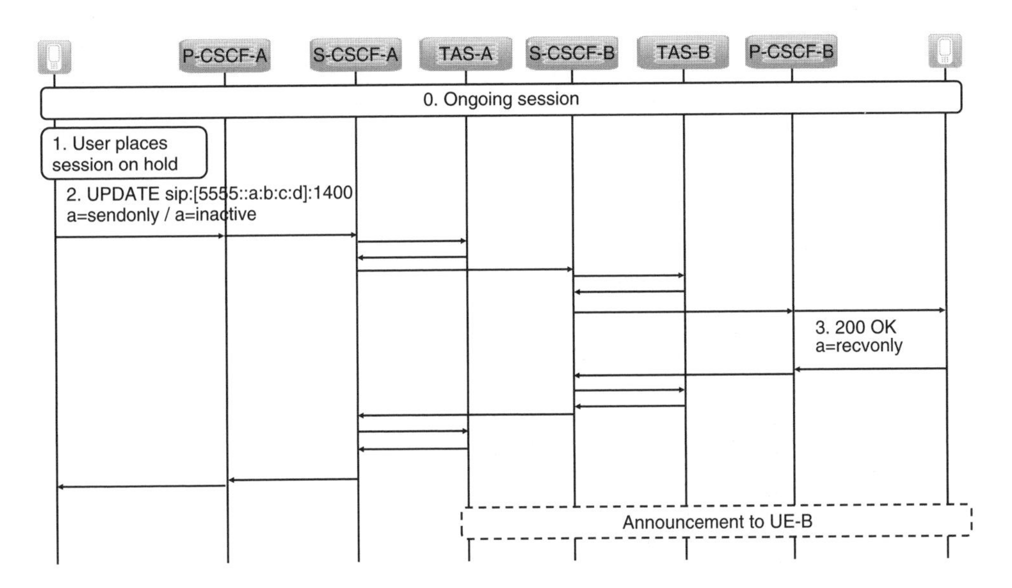

Figure 4.27 Example of communication hold supplementary service.

later time. When a user wishes to place communication on hold the UE sends either an UPDATE or a RE-INVITE request towards to the other user. The attribute line in the SDP of the request having the value 'inactive' or 'sendonly' indicates that the sender is not willing to receive media stream(s) from the other user, as shown in Figure 4.27. When a user wished to resume held media stream(s) the UE sends an UPDATE or a RE-INVITE request towards the other user. The attribute line in the SDP of the request having the value 'sendrecv' or 'recvonly' indicates that the sender is again willing to receive media stream(s) from the other user. The TAS of the served user may provide an announcement to the other user.

4.6.3.6 Conference

Conference service enables a user to communicate with one or more users simultaneously. This supplementary service re-uses a solution developed for standalone IMS conferencing. Figure 4.28 shows an example of how a user can transform ongoing 1 : 1 sessions to multimedia conference. This can be seen as an equal service to existing GSM supplementary services known as multiparty (MPTY) conference. To create an ad-hoc conference user A takes action and the UE of this user will create an ad-hoc conference by sending an INVITE to SIP-URI known as conference-factory URI (e.g. sip:conferencefactory@example.com). On reception of such an INVITE the TAS creates an ad-hoc conference and returns conference identifier, conference URI, in the SIP response. Once the UE A receives the conference URI it will send one REFER to the conference URI (TAS) per user that it wants to get invited to the conference. The REFER request contains instruction to invite particular user (B or C) to this newly created conference. The request will also contain instructions towards the UE B/UE C that new incoming request is supposed to replace the existing session between UE A and UE B/UE C.

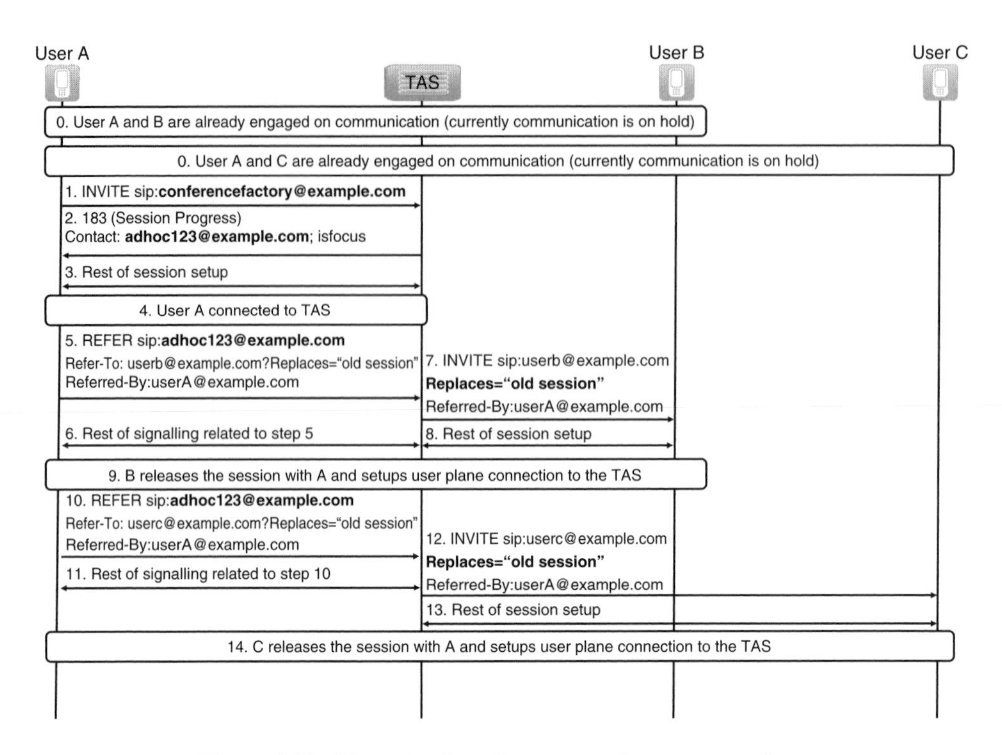

Figure 4.28 Example of conference supplementary service.

The REFER request gets routed using normal IMS routing principles to an AS providing the conferencing functionality. When the AS gets this request it forms two independent session requests towards UE B and UE C. Once the UE B/UEC receives a new INVITE it learns that this is a request to join a conference and to release the existing session with UE A. After additional SIP signalling and user-plane setup users A, B, C have joined the conference and can start discussing.

Here a subset of IMS conference functionality was described as required for MPTY. It is important to acknowledge that IMS MMTel conference service does not require an existing IMS MMTel session prior to conference, so this is a major differentiator compared to the existing GSM service.

4.6.3.7 Communication Diversion

The CDIV (aka communication forwarding) service allows the user to re-direct an incoming request that fulfils certain provisioned or configured conditions to another destination. The baseline for this service is inherited from established diversion services in public switched telephone network (PSTN)/integrated services digital network (ISDN) networks. Namely the following services are included in the GSM service set:

- Communication forwarding unconditional (CFU) service enables a served user to have the network redirect to another user communications which are addressed to the served user's address.

- Communication forwarding busy (CFB) service enables a served user to have the network redirect to another user communications which are addressed to the served user's address while the served user is busy.
- Communication forwarding no reply (CFNR) service enables a served user to have the network redirect to another user communications which are addressed to the served user's address, and for which the communication is not answered within a defined period of time.
- Communication forwarding on not logged-in (CFNL) service enables a served user to redirect incoming communications which are addressed to the served user's address, to another user (forwarded-to address) in case the served user is not registered (logged-in).
- Communication diversion on mobile subscriber not reachable (CFNRc) service enables a user to have the network redirect all incoming communications, when the user is not reachable (e.g. there is no IP connectivity to the user's terminal), to another user.
- Communication deflection (CD) service enables the served user to respond to an incoming communication by requesting redirection of that communication to another user (before ringing and after ringing).[5]

In addition to previously listed conditions the following new conditions are defined [support of these advanced conditions and the communication diversion notification service are not required in GSMA (IR.92)]:

- Communication forwarding depending on called user's presence status;
- Communication forwarding depending on the calling user's identity or lack of identity;
- Communication forwarding depending on media including in the incoming session;
- Communication forwarding depending on time of the call.

For example the user could create rules to divert incoming audio call to CS domain, divert incoming video call when user's presence status is 'silent' to video mail box, divert MSRP session to client associated to fixed device, divert unanswered session to a third party if no answer within 20 s. These and a great number of other possibilities are enabled by standards and the user is able to configure these settings using the XCAP protocol, as described in Section 4.6.3.11. Once the rules are set then the TAS will enforce communication according to the user's preferences.

Figure 4.29 shows an example when user B wants to divert all incoming communication to user C (+358501234567). First the device uses XCAP protocol to set up diverting rules to the TAS. Once the TAS serving user B receives an incoming communication attempt it checks if the served user has uploaded any settings for an incoming session treatment and based on the stored information it makes a decision to divert the request to user C. The sent request to user C will contain additional information showing that the call was originally targeted to user B and it additionally conveys the reason for diversion (these are carried in the History-Info header). Moreover the TAS informs the caller that the communication is being forwarded (SIP response 181 Call is Being Forwarded).

[5] Communication deflection is not required in GSMA (IR.92).

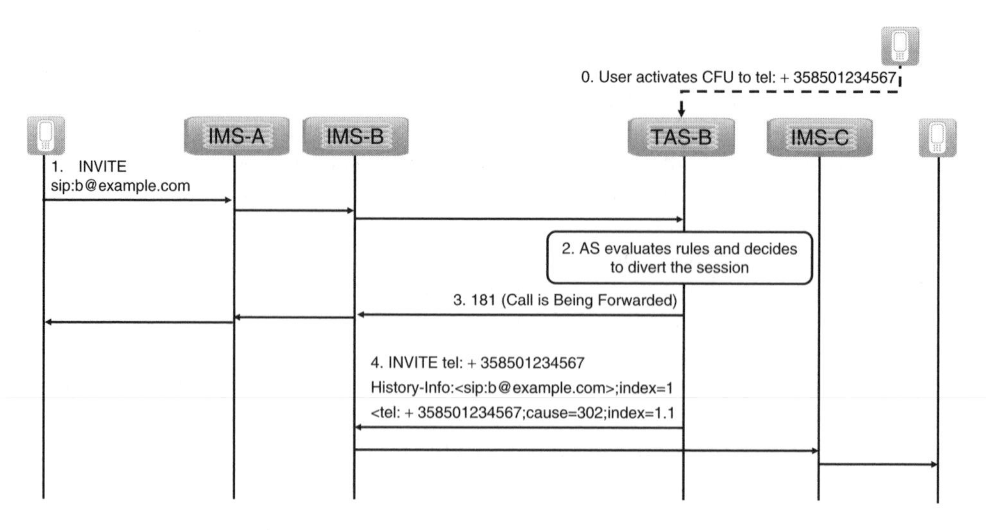

Figure 4.29 Example of communication diversion supplementary service.

4.6.3.8 Communication Barring

The communication barring service allows users to selectively block session attempts. The barring service can be further divided into three main classes: incoming communications barring (ICB), outgoing communication barring (OCB) and anonymous communication rejection (ACR). The ICB is a service that rejects incoming communications that fulfil certain provisioned or configured conditions on behalf of the terminating user. The ACR is a particular case of the ICB service, that allows barring of incoming communications coming from an anonymous originator. The OCB is a service that rejects outgoing communications that fulfil certain provisioned or configured conditions on behalf of the originating user. For example, the user can bar originating video call attempts while roaming, bar originating phone calls to specific numbers between 8 a.m. and 4 p.m., bar anonymous terminating messaging sessions attempts or bar incoming calls from numbers +358401234567, +358501234567 and allow the rest of the calls. These and a great number of other conditions and rules have been standardised and the user is able to configure these settings using the XCAP protocol, as described in Section 4.6.3.11. Once the rules are set then the TAS will enforce communication according to the user's preferences.

GSMA (IR.92) requirements for the communication barring service are inherited from the current GSM speech service. Therefore the only conditions to provide the following services are required: barring of all incoming calls, barring of all outgoing calls, barring of outgoing international calls, barring of outgoing international calls – ex home country and barring of incoming calls – when roaming.

In Figure 4.30 user B uses the XCAP protocol to bar all incoming communication when roaming. When user B later receives a communication attempt the TAS providing service for her utilises the stored information and, based on knowledge of the roaming status of user B, it blocks the request with the appropriate SIP error response, 603 Decline, which means that the user explicitly does not wish to take part in communication.

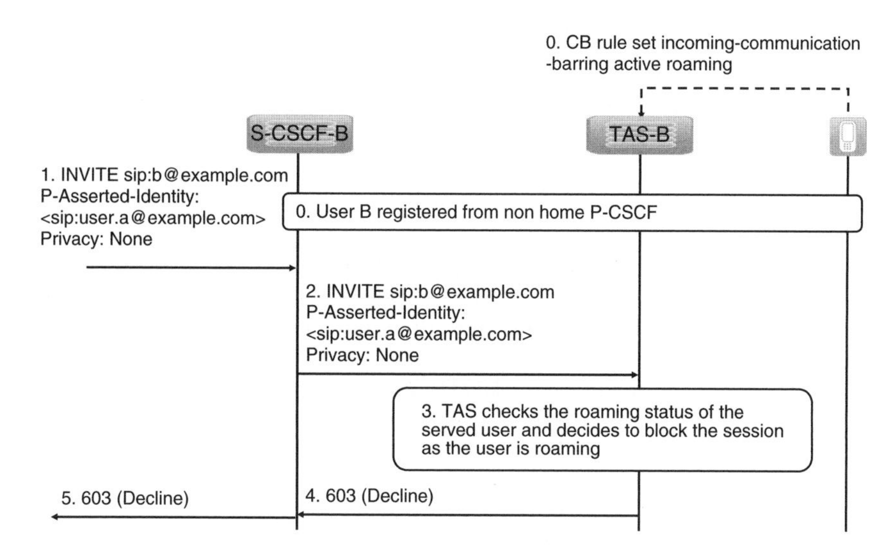

Figure 4.30 Example of incoming communication barring supplementary service.

4.6.3.9 Message Waiting

The Message waiting indication (MWI) service enables the TAS to indicate to the user that there is at least one message waiting in the message account. To activate this service the UE sends SIP SUBSCRIBE to the network having suitable value in the Expire header, for example one day.

```
SUBSCRIBE sip:kristiina@example.com SIP/2.0
Via: SIP/2.0/UDP [5555::a:b:c:d]:1400;branch=4uetb
Route: <sip:[5555::55:66:77:88];lr>
Route: <sip:orig@scscf1.home.fi;lr>
From: <kristiina@example.com>;tag=7547d
To: <kristiina@example.com>
Event: message-summary
Expires: 86400
Accept: application/simple-message-summary
Contact: <sip:[5555::a:b:c:d]:1400>
Content-Length: 0
```

The subscription is intended for an event named 'message-summary', that uniquely defines that the user is interested in receiving MWIs; this is identified in the Event header of the request.

The Request-URI identifies the user whose message account information is requested and therefore has to be set to user's public user identity or public service identity of the message account depending on the communication service provider's policy on how to access the MWI service and must also be indicated in the to header.

The Accept header indicates that only information of the type 'application/simple-message-summary' can be processed by the UE for this subscription, which is a simple text-based format for MWIs.

The AS like TAS receives this SUBCRIBE request and will check whether the requesting user is allowed to subscribe to this particular message account. As Kristiina is subscribing to her own account in this case, this is allowed. Therefore, the AS will immediately:

- Return a 200 (OK) response for the SUBSCRIBE request, indicating that the subscription was successful.
- Generate information regarding the current state of the message account. This initial information only contains a summary of message account. Further notifications may contain extended information (e.g. the most important headers of the message: to, from, subject, date, priority).
- Send the generated information in a NOTIFY message towards the user (in this case Kristiina's UE).

```
NOTIFY sip:kristiina@example.com SIP/2.0
From: <sip:kristiina@example.com>;tag=31415
To: <sip:kristiina@example.com>; tag=7547d
Subscription-State: active; expires=86399
Event: message-summary
Content-Type: application/simple-message-summary
Content-Length: (...)
Messages-Waiting: yes
Message-Account: sip:kristiina@example.com
Voice-Message: 2/1 (1/0)
Video-Message: 0/1 (0/0)
```

Not all the information that is included in the NOTIFY request is shown here – the above headers are only those that are necessary to understand the nature of the message-summary event.

In this example it is assumed that the message account at the moment of subscription has three voice messages (two new and one old, with one new message being urgent) and one old video message.

This solution is based on the SIP event framework as defined in (RFC3265) and the event package defined for MWI (RFC3842). 3GPP has defined some restrictions in the MWI specification (3GPP TS 24.606) compared to (RFC3842). (3GPP TS 24.606) supports limited coding of the message types (voice-message, video-message, fax-message, pager-message, multimedia-message, text-message, none) and limited information of stored messages (e.g. to, from, subject, date, priority) (3GPP TS 24.606).

It should be noted that 3GPP has not standardised how to fetch actual messages. For example when retrieving voice/video messages a user could dial her voice/video mail number.

4.6.3.10 Communication Waiting

The general service characteristic for communication waiting (CW) is that it allows a user that has activated the service and is currently having an active call to be informed

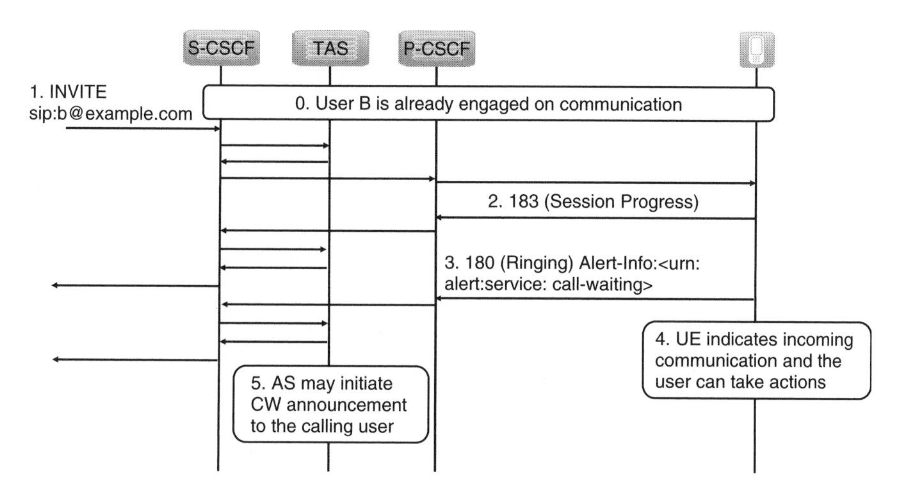

Figure 4.31 Example of communication waiting supplementary service.

about an additional incoming call. For CW there is a network- and a UE-centric mode. In network-centric mode an AS takes care of determining the availability of resources for an additional call, while in UE-centric mode the UE of the served user will decide on its own whether or not to treat an additional incoming call as a waiting call. In both cases the served user gets an indication of the waiting call, for example via a tone or a visual signal. For CW the user has the possibility to manage the service via XCAP over Ut interface which allows activation and deactivation of the network centric mode. GSMA (IR.92) endorsed the UE centric mode.[6] Figure 4.31 depicts how it works.

When a call is treated as a waiting call then the originator of the call will be notified about this situation. To achieve this, the 180 (RINGING) response sent back to the originating side will contain an Alert-Info header field set to '<urn:alert:service:call-waiting>' as defined in (draft-ietf-salud-alert-info-urns). In addition the TAS serving the user B may play an announcement to the calling user 'please hold the user you have dialled is engaged in another call'.

4.6.3.11 Supplementary Services Management

Supplementary services management allows the user to manage the supplementary services configuration. It uses extensible markup language (XML) configuration access protocol (XCAP) as its underlying mechanisms.

With XCAP a user is able to upload information to an XCAP server, which provides this uploaded information to ASs that use this information to satisfy a request demanded by the user. With XCAP, the user is also allowed to manipulate, add and delete such data. XCAP uses the hypertext transfer protocol (HTTP) to upload and read the information set by users and the information is represented using XML. Applications need to define an XCAP application usage ID (AUID), which defines the way that a unique application can make use of XCAP. It defines the XML document for the application. There are four

[6] In UE centric mode XCAP configuration is not required.

operations inherited from HTTP: create (HTTP put), fetch (HTTP get), modify (HTTP put) and delete (HTTP delete). Each of these operations can be targeted to the whole document, an element in the document or an attribute in the document.

XCAP usage for supplementary services was invented in ETSI TISPAN around 2005 and it was included in ETSI TISPAN NGN Release 1.

The supplementary service settings are stored as an XML document called the simservs document. This document defines an authorisation policy as a set of rules – called a 'ruleset' which describes the handling of SIP requests in order to provide supplementary services to the served user. Each rule grants permission based on certain matching criteria, like the called user does not answer. These matching criteria are called 'conditions'. Permission further defines an action. The action part dictates in which way the system should act, for example forward the incoming session to a voice mail.

Several supplementary services can be configured with XCAP. Here we focus on those services which are included in GSMA (IR.92) and are configurable using XCAP namely CDIV and communication barring.[7]

Communication Diversion

To enable this service a 'ruleset' has been defined. The following conditions have been defined: the called user is busy, the called user is not registered, the called user's current presence status, the identity of the calling user, anonymous calling user, time, offered media, the called user does not answer, an identity of the calling user derived from an external list, other identities that are not present in any rule, the called user is not reachable. When the condition or set of conditions is evaluated true then the only defined action is to forward the session to a given destination. An example of authorisation policies: if I am not responding then forward the session to multimedia mailbox; if I am busy and my presence status is meeting and the caller is my wife then forward to my secretary; if I am not registered to the network then forward to multimedia mailbox.

GSMA (IR.92) only requires support of following five conditions: busy, media, no-answer, not-registered and not-reachable.

- Busy evaluates to true when the called user is busy.
- Media evaluates to true when the incoming call request for certain media (supported media types: audio, audio and video).
- No-answer evaluates to true when the called user does not answer (no-answer timeout is detected or when a no answer indication is received).
- Not-registered evaluates to true when the called user is not registered.
- Not-reachable evaluates to true when there is a signalling channel outage during session setup to the served user's UE and the served user is registered.

[7] Other services which can be managed with XCAP are: OIR, TIP, CCBS, CCNR, FA, CUG and CW.

These conditions are sufficient to support present GSM call forwarding supplementary services as follows:

Ruleset and condition = true	Decision in TAS
Communication diversion activated and no specific condition included	CFU executed
Busy	CFB executed
No-answer	CFNR executed
Not-registered	CFNL executed
Not-reachable	CFNRc executed

Communication Barring

To enable this service a 'ruleset' has been defined. The following conditions have been defined: the identity of the calling user, the identity of the called user, an identity of the calling user derived from an external list, other identities that are not present in any rule, anonymous calling user, offered media, the called user's current presence status, time, diverted session, roaming status, calling to other country, calling from other country except while calling to home country. If all conditions in a rule are true then the rule matches and the specified action is executed. Action takes either value true or false. When one of the actions is evaluated true then the session is allowed to continue otherwise the session is blocked.

GSMA (IR.92) only requires support of following three conditions: roaming, international and international-exHC.

- Roaming evaluates to true when the served user is registered from an access network other then the served users home network.
- International evaluates to true when the request URI of the outgoing SIP request: corresponds to a telephone number and does not point to a destination served by a network within the country where the originating user is located when initiating the call.
- International-exHC (international barring, excluding the home country) evaluates to true when the request URI of the outgoing SIP request:
  - Corresponds to a telephone number;
  - Does not point to a destination served by a network within the country where the originating user is located when initiating the call;
  - Does not point to a destination served within the served users home network.

These conditions are sufficient to support present GSM call barring supplementary services as follows:

Ruleset	Action	Decision in TAS
Incoming barring activated and no specific condition included	False	Barring of all incoming calls
Outgoing barring activated and no specific condition included	False	Barring of all outgoing calls
International and outgoing barring activated	False	Barring of outgoing international calls
International-exHC and outgoing barring activated	False	Barring of outgoing international calls – ex home country
Roaming and incoming barring activated	False	Barring of incoming calls – when roaming

Examples of more advanced authorisation policies:

Ruleset	Action	Decision in TAS
Session is received from Bob	False	Session is blocked
Session is not coming from Robert	True	Session is allowed
If the request is for video	False	Session is blocked
My presence status is offline and time is 8 a.m. to 4 p.m.	False	Session is blocked

4.6.3.12 Multimedia Telephony Service Management Example

Figure 4.32 shows an example what happens when a user wants to store her communication diversion (unconditional) setting to the network.

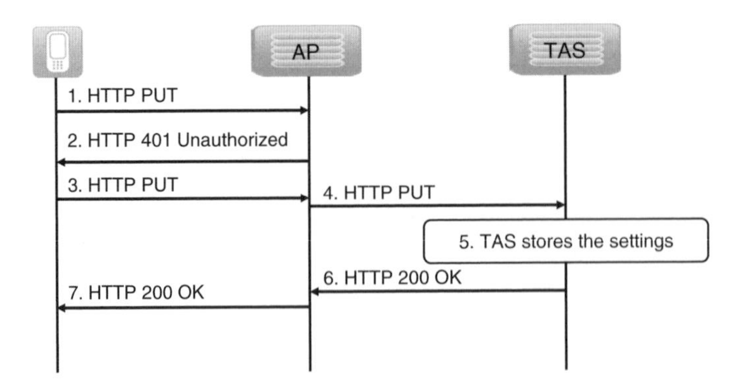

Figure 4.32 Multimedia telephony service management example.

UE creates a XCAP request and sends it to the network (step 1). The AP in the path detects unauthorised request and it requests authorisation (step 2). The UE repeats the request including the Authorisation header containing the required information (step 3).

```
PUT /simservs.ngn.etsi.org/users/sip:user1@home1.net/simservs/~~/
     communication-diversion HTTP/1.1
Host: xcap.mnc091.mcc244.ipxuni.3gppnetwork.org
Date: Thu, 25 Aug 2011 11:04:53 GMT
Authorisation: Digest realm="xcap.mnc091.mcc244.ipxuni.
     3gppnetwork.org", nonce="e966c32a924255e42c8ee20ce7f6",
     username="kristiina@example.com", qop=auth-int,
uri="/simservs.ngn.etsi.org/users/sip:kristiina@example.com/
     simservs/~~/communication-diversion/ruleset", response=
     "adq3283hww88whhjw98822333ddd32", cnonce=
     "wqesatt874873j3gg3kk39944hhhee", nc=00000001


X-3GPP-Intended-Identity: sip:kristiina@example.com
Content-Type: application/xcap-el+xml; charset="utf-8"
Content-Length: (...)
    <communication-diversion active="true">
     <cp:ruleset>
      <cp:rule id="rule1">
        <cp:conditions>
        </cp:conditions>
        <cp:actions>
          <forward-to>
            <target>tel:+358405556667</target>
            <notify-caller>true</notify-caller>
          </forward-to>
        </cp:actions>
      </cp:rule>
    </cp:ruleset>
   </communication-diversion>
```

This time the AP conducts successful authorisation and forwards the request to TAS (step 4). The TAS validates the received input and stores the supplementary service data (step 5) and accepts the settings (steps 6 and 7).

5

VoLTE End to End and Signalling

5.1 Overview

This chapter gives a detailed example of voice over long term evolution (VoLTE) related procedures in end to end manner. Signalling procedures, protocol messages and their elements are described and explained based on an example. The reader will see how VoLTE signalling works and how previously described concepts and architecture are realised at the protocol level. Nevertheless, this part concentrates on straightforward examples and does not handle error or abnormal procedures in detail. To give a better understanding of the procedures applied, each section of this chapter concentrates on different concepts, such as VoLTE subscription, evolved packet system (EPS) attach and default bearer, voice domain discovery procedure, IP multimedia subsystem (IMS) registration, IMS voice over Internet protocol (VoIP) session, circuit switched fallback (CSFB) procedures, emergency session and messaging.

This chapter is based on the assumption that user Kristiina has just obtained a new long term evolution (LTE) mobile and wants to call her friend Jennifer who is living in the United States. Kristiina's communication service provider is in Finland and it offers VoLTE as the primary voice service but has not yet completed a full roll-out of LTE coverage. For securing end user satisfaction the communication service provider has also deployed a single radio voice call continuity (SRVCC) feature and IMS centralised services (ICS) to enable secure seamless service continuity when a VoLTE subscriber goes outside of the LTE coverage. We will introduce what kind of user profile (Section 5.2) the Finnish communication service provider has created for Kristiina. Furthermore we will explain how Kristiina's user equipment (UE) performs EPS attach (Section 5.3) and discovers whether IMS VoIP is to be used instead of CSFB for voice communication (Sections 5.3 and 4.3.2.3). Section 5.4 contains an IMS registration walk through. In Section 5.5 we will show what happens when Kristiina makes an IMS VoIP session to her friend Jennifer. Here we will reveal session initiation protocol (SIP) session setup with relevant SIP details and related policy and charging control (PCC) interactions to setup dedicated bearer and associated guaranteed bit rate (GBR) radio bearer. Section 5.6 provides a description of what happens when Kristiina moves to the edge of LTE coverage

---

Voice over LTE: VoLTE, First Edition. Miikka Poikselkä, Harri Holma, Jukka Hongisto, Juha Kallio and Antti Toskala.

and network seamlessly moves ongoing IMS VoIP call to a 2G/3G circuit switched (CS) call. In Section 5.7 we will show a special session case, an IMS emergency session. Section 5.8 explains what CSFB is and how it works. LTE messaging solutions complete this chapter in Section 5.9.

5.2 VoLTE Subscription and Device Configuration

At first the communication service provider needs to create a VoLTE subscription for Kristiina. This subscription is permanently stored in the home subscriber server (HSS) and necessary information of it is downloaded to a mobility management entity (MME) when the user performs EPS attach (see Section 5.3) and to a serving call session control function (S-CSCF) when the user performs IMS registration (see Section 5.4.7). Figure 5.1 shows the evolved packet core (EPC) part of the VoLTE subscription data. Table 5.1 explains the purpose of each EPC parameter and gives example value of the parameter for VoLTE usage. Figure 5.2 shows the IMS part of the VoLTE subscription data. Table 5.2 explains the purpose of each IMS parameter and gives example value of the parameter for VoLTE usage.

Once the communication service provider has created the subscription record it can use device management to provision Kristiina's UE with the communication service provider voice domain preferences (3GPP TS 24.167). The preferences define how the UE that is both CSFB and IMS capable is supposed to handle VoLTE communication. Four different values exist, as depicted in Table 5.3 (3GPP TS 23.221).

Hence we assume that VoLTE UE is configured as 'IMS PS voice preferred, CS voice as secondary'. Furthermore we assume that the VoLTE UE behaves as voice-centric (defined

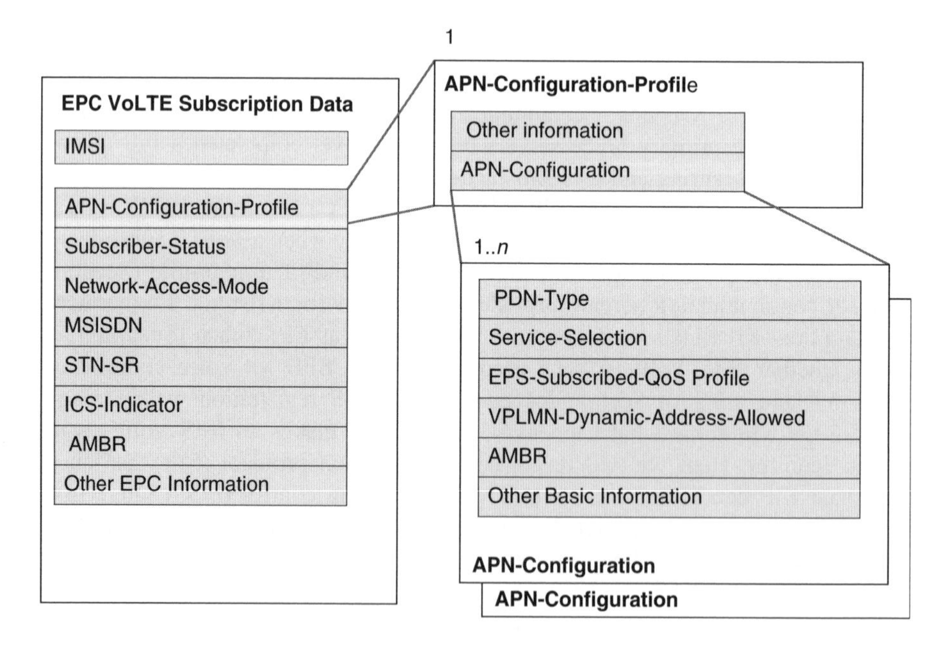

Figure 5.1 EPC VoLTE subscription information.

Table 5.1 EPC VoLTE subscription parameters and example values

Parameter	Purpose	Example value/content
IMSI	Unique subscription identifier, which is used, for example to retrieve subscription data from HSS	310150987654321
APN-configuration-profile	APN configurations for EPS, with one APN-configuration item marked as a default APN configuration	See Figure 5.1
APN-configuration	Subscribed APN configurations such as IMS APN	See Figure 5.1
PDP-type	Address type of PDN: IPv4 (0), IPv6 (1), both in IPv4 mode and in IPv6 mode (dualstack usage) (2), IPv4 or IPv6 (3)	2 (IPv4IPv6)
Service-selection	APN network identifier or contain the wild card value	IMS
EPS-subscribed-QoS profile	Bearer-level QoS parameters (QoS class identifier and allocation retention priority) associated to the default bearer for an APN	5 (QCI5), ARP9
VPLMN-dynamic-address-allowed	Visited PDN GW usage	1 (allowed)
AMBR	Aggregate maximum rit rate (AMBR) uplink/downlink per subscription/APN	500 000 1 000 000 (uplink and downlink separately)
Subscriber-status	Service is granted or barred	0 (service granted)
Network-access-mode	To indicate the network-access-mode AVP is of type enumerated, with the following defined values: PACKET_AND_CIRCUIT (0), Reserved (1), ONLY_PACKET (2)	0 (packet and circuit services enabled)
MSISDN	Number which uniquely defines the mobile subscriber at international level	358501234567
STN-SR	Identifies the session transfer number for SRVCC Note: the removal of this parameter can deactivate SRVCC service in the serving PLMN	35850987654321
ICS-indicator	ICS service is provisioned or not	True

by manufacturer or end user). Voice-centric means that the UE always try to ensure that a voice service is possible (3GPP TS 23.221). In other words if voice-centric UE discovers (see Section 5.3) that IMS voice and CSFB are not supported then it needs to disable LTE capability and re-select other radio access such GSM/EDGE radio access network (GERAN) or UMTS terrestrial radio access network (UTRAN). In addition, the communication service provider can define its preference for the domain to be used for a short message service (SMS) (SMS-Over-IP is used or not).

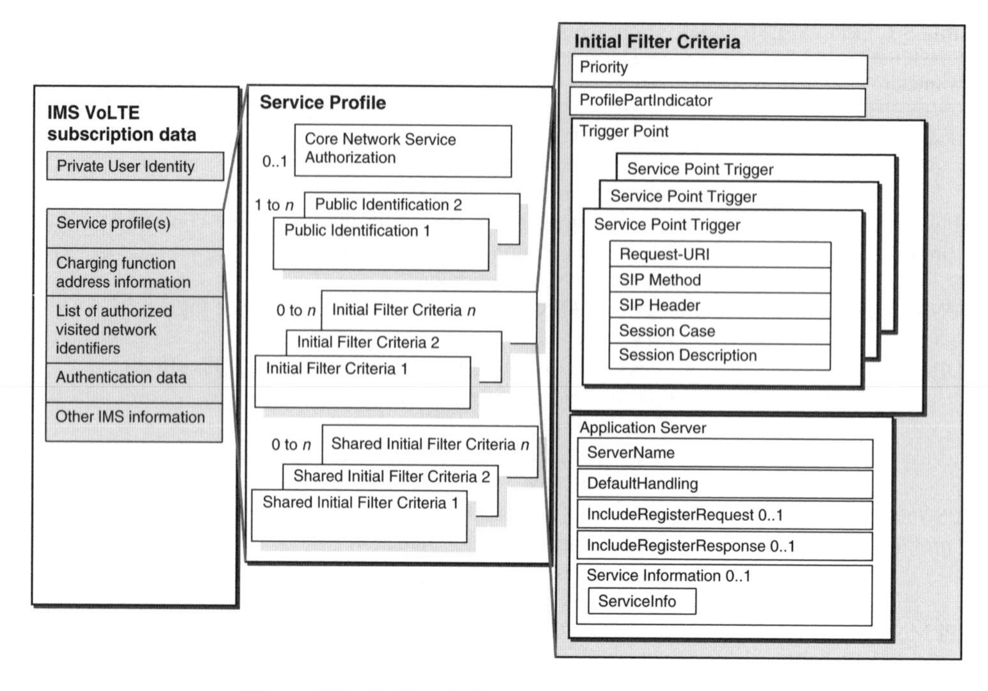

Figure 5.2 IMS VoLTE subscription information.

Base standards:

3GPP TS 23.003	Numbering, addressing and identification
3GPP TS 23.008	Organisation of subscriber data
3GPP TS 23.221	Architectural requirements
3GPP TS 24.167	3GPP IMS management object (MO)
3GPP TS 29.272	Evolved packet system (EPS); mobility management entity (MME) and serving GPRS support node (SGSN) related interfaces based on diameter protocol
3GPP TS 29.228	IP multimedia (IM) subsystem Cx and Dx interfaces; signalling flows and message contents
3GPP TS 29.229	Cx and Dx interfaces based on the diameter protocol; protocol details

5.3 EPS Attach for CSFB/IMS VoIP and Default Bearer Activation

This section describes how a user makes an initial attachment to the LTE network, which provides IP connectivity for IMS services. The EPS attachment is done when the user switches UE on in available LTE access (Figure 5.3). Here we will now focus on a

Table 5.2 IMS VoLTE subscription parameters and example values

Parameter	Purpose	Example value/content
Private user identity	Unique subscription identifier, which identifies the user from the IMS network perspective	private_user1@example.com
Charging function address information	Primary and secondary addresses for offline/online charging entity	aaa://chargingentity.example .com
List of authorised visited network identifiers	The list of visited network identifiers which are allowed for roaming	csp1.net,csp2.com,csp99.com, and so on
Authentication data	Contains all required information for IMS authentication	Authentication schema (IMS AKA), shared secret (needed for to create random challenge, expected result (of random challenge), confidentiality key, integrity key, network authentication token) and sequence numbers
Other basic IMS information	Such as associated IMS private user identities (IMPIs), associated registered IMPIs, loose-route indication, S-CSCF restoration, and so on	
Service profile	Collection of user specific information	See Figure 5.2
Core network service authorisation	Contains media policy and IMS communication service identifier policy	
	Media policy information contains an integer that identifies a subscribed media profile in the S-CSCF	2 (media policy)
	The IMS communication service identifier policy contains a list of service identifiers that identifies which IMS communication service user is entitled to use	urn:urn-7:3gpp-service.ims.icsi.mmtel (allowed IMS communication service identifier)
Public user identity	User identity in IMS network, used for requesting communication with other users	sip:kristiina@example.com and tel: +358 50 1234567

(*continued overleaf*)

Table 5.2 *(continued)*

Parameter	Purpose	Example value/content
Initial filter criteria	User specific service triggering information	See Figure 5.2 and Section 4.5
Shared initial filter criteria	Common service triggering information for multiple users which contains a reference value to initial filter criteria which are locally administrated and stored in S-CSCF	1 (for example identifies initial filter criteria for IMS multimedia telephony)

Table 5.3 VoLTE UE and communication service provider's preferences for voice service

Configuration value	Purpose
CS voice only	UE will use only the CS domain to originate voice calls and it will not attempt to initiate voice sessions over IMS using a PS bearer.
CS voice preferred, IMS PS voice as secondary	If CS voice is available the UE will use the CS domain to originate and terminate voice calls.
IMS PS voice preferred, CS voice as secondary	If IMS voice is available the UE will use IMS to originate and terminate voice sessions.
IMS PS voice only	The UE will use IMS to originate voice sessions and will not attempt CS voice.

voice-centric device that is willing to use IMS VoIP. See Section 5.8 for CS fallback and Section 5.9.2.1 for SMS over SGs.

1. Before initial attachment the UE makes a radio connection setup.
2. The UE initiates the attach procedure by sending an attach request to the eNodeB. The message contains UE voice capabilities and preferences, for example:
   a. Voice domain preference for Evolved UMTS terrestrial radio access network (E-UTRAN) = IMS PS voice preferred, CS voice as secondary;
   b. UE's usage setting = voice-centric;
   c. SRVCC to GERAN/UTRAN capability = SRVCC from UTRAN high-speed packet access (HSPA) or E-UTRAN to GERAN/UTRAN supported.
3. The eNodeB forwards the attach request message to the selected MME in a S1-MME control message.
4. MME initiates this procedure for authenticating the UE and involves HSS to obtain authentication vectors for the UE.

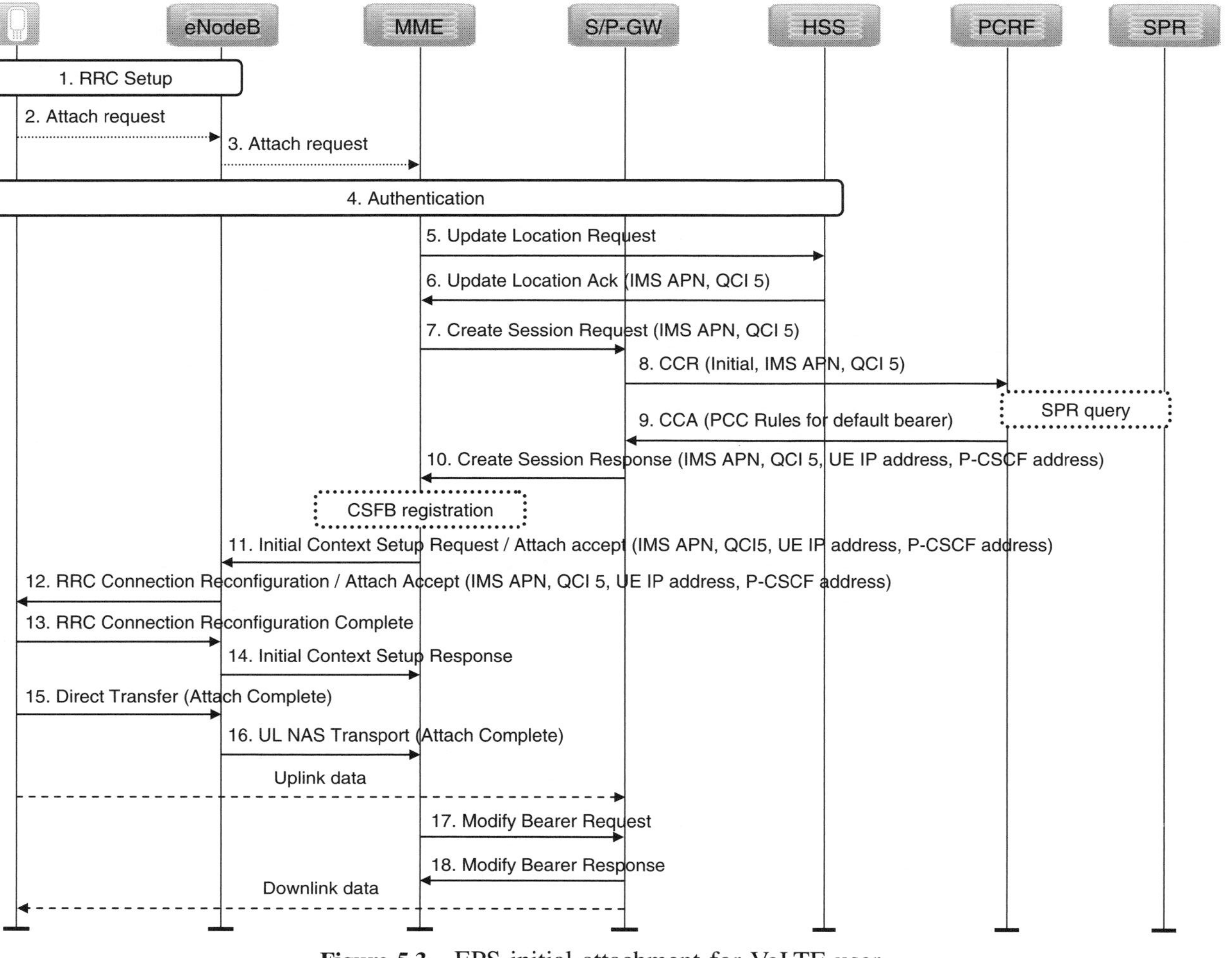

Figure 5.3 EPS initial attachment for VoLTE user.

5. MME sends an update location request message to the HSS to update the records for the given international mobile subscriber identifier (IMSI).
6. The HSS acknowledges the update location message by sending an update location acknowledgement message to the MME. The default *packet data network (PDN) subscription context* in this example IMS access point name (APN) and quality of service (QoS) parameters used for SIP (XML configuration access protocol) signalling bearer.
   - PDN type = IPv4v6;
   - Access point name = IMS;
   - QCI = 5;
   - Allocation and retention priority (ARP) priority level = 9;
   - ARP PRE-EMPTION_CAPABILITY_DISABLED = 1;
   - ARP PRE-EMPTION_VULNERABILITY_ENABLED = 0;
   - Subscribed-APN-aggregate maximum bit rate (AMBR) = 500 000 (UL), 1 000 000 (downlink, DL);
   - Visited public land mobile network (VPLMN) address allowed = 1 (allowed);

   VoLTE related HSS data is further explained in Section 5.2.
7. The MME validates the UE voice capabilities, subscriber's APN and QoS parameters based on the home location register (HLR)/HSS subscription data and possible operator configurations. For example there might be restrictions for the roaming subscribers. For the VoLTE subscribers it is preferred that *IMS* APN is used as the default APN. The MME selects an serving gateway (S-GW) and PDN gateway (P-GW). The S-GW is selected based on the, for example UE location, and the P-GW is selected based on the APN. For voice users there can be dedicated P-GW(s), which support *IMS* APN. The 'ims P-GW' may locate in the same operating site where media gateways (MGWs) are located. After S/P-GW selection the MME sends a create session request message to the selected S-GW to initiate the creation of default bearers for the UE, which is forwarded to P-GW.

8-9. For the IMS voice user there is always a dynamic PCC deployed and the P-GW performs IP connectivity access network (CAN) session establishment procedure to obtain the default PCC rules for the UE (see also Section 4.3.2.2. The request to the policy and charging rules function (PCRF) contains at least the user identification, UEs Ipv4 address and/or Ipv6 prefix, IP-CAN type and radio access type. The PCRF may request subscriber specific information from the subscription profile repository (SPR), or the PCRF may use the same rule set for all subscribers, for example on an APN basis. The PCRF may provide PCC rules, bearer authorisation and QoS parameters for the IMS PDN connection including a maximum bit rate, QoS class identifier (QCI) and ARP.

10. The PDN-GW creates a new entry in its EPS bearer context table for user plane routing between the S-GW and PDN. The P-GW responds to the S-GW with a create session response message and S-GW responds to the MME with the create session response message including default bearer identification, EPS bearer QoS, UE IP address and P-CSCF address. The P-CSCF address (for example 5555::55:66:77:88) is sent in the protocol configuration options (PCOs) inside the create session response message.

11. MME checks any modifications for the default bearer setup, calculates the UE-AMBR based on the APN-AMBR. If the UE has made a combined EPS/IMSI attach in step 2 then it also makes IMSI attach to the MSC server (MSS) for CSFB service, then MME needs to derive the visitor location register (VLR) address and MME sends a location update request to the mobile switching centre (MSC)/VLR and continues the attach procedure after location update accept. See Section 5.8.5 for CSFB registration.
    The MME responds to the UE with an attach accept message via eNodeB. The MME makes also a final validation *IMS voice over PS session* indication for the VoLTE user (see also Section 4.3.2.3) and the EPS bearer parameters, which are sent by network to UE, can be for example:
    a. PDN type = IPv4v6;
    b. PDN address = 192.168.10.10 (IPv4), 5555::a:b (/64 IPv6 prefix);
    c. APN = ims;
    d. QCI = 5;
    e. APN-AMBR = 500 000 (UL), 1 000 000 (DL);
    f. IMS voice over PS session = 1 (supported);
    g. Proxy-call session control function (P-CSCF) address: 5555::55:66:77:88.
    UE gets also tracking area identity (TAI) list, which is the area where UE can move without any mobility update towards network. *IMS voice over PS supported* indication is valid for that area, but may be different when UE moves out of this area.
12. The eNodeB makes radio resource control (RRC) reconfiguration. UE gets bearer information, EPS bearer QoS, UE IP address and P-CSCF address needed for the IMS SIP signalling.
13. RRC reconfiguration is completed.
14. The eNodeB sends the initial context response message to the MME.
15. The UE sends a direct transfer message to the eNodeB, which includes the attach complete (EPS bearer identity, non access stratum (NAS) sequence number (SQN), NAS–Medium Access Control (MAC)) message.
16. The eNodeB forwards the attach complete message to the new MME in an uplink NAS transport message. The UE can now send uplink packets towards the eNodeB which will then be tunnelled to the S-GW and P-GW.
17. MME sends a modify bearer request message to the S-GW.
18. The S-GW acknowledges by sending modify bearer response to the MME. Bearers are now established and ready to exchange uplink and DL packets.

5.4 IMS Registration

The following procedures are performed during IMS registration (see Figure 5.4):

- The UE sends a REGISTER message to home network to perform SIP registration for public user identity.
- The interrogating-call session control function (I-CSCF) selects the S-CSCF that serves the user while it is registered.

- The S-CSCF downloads the authentication data of the user from the HSS, and the HSS performs an IMS roaming check.
- The UE and the P-CSCF agree on a security mechanism.
- The UE and the network (S-CSCF) authenticate each other.
- IP Security (IPSec) associations between the UE and the P-CSCF are established.
- SIP compression may start between the UE and the P-CSCF (SIP compression is not used in LTE access and it is not described here).
- The UE learns the route to the S-CSCF.
- The S-CSCF learns the route to the UE.
- The S-CSCF downloads the user profile of the user from the HSS.
- The S-CSCF registers the default public user identity of the user.
- The UE registers the supported IMS communication services at the S-CSCF.
- The S-CSCF may, based on the user profile, implicitly register further public user identities of the user.
- The S-CSCF may, based on the user profile, perform registration to application servers (ASs) on behalf of the user.
- The UE becomes aware of all public user identities that are assigned to user and her current registration state.
- The P-CSCF becomes aware of all public user identities that are assigned to user and her current registration state.
- The P-CSCF becomes aware of charging function address(es).

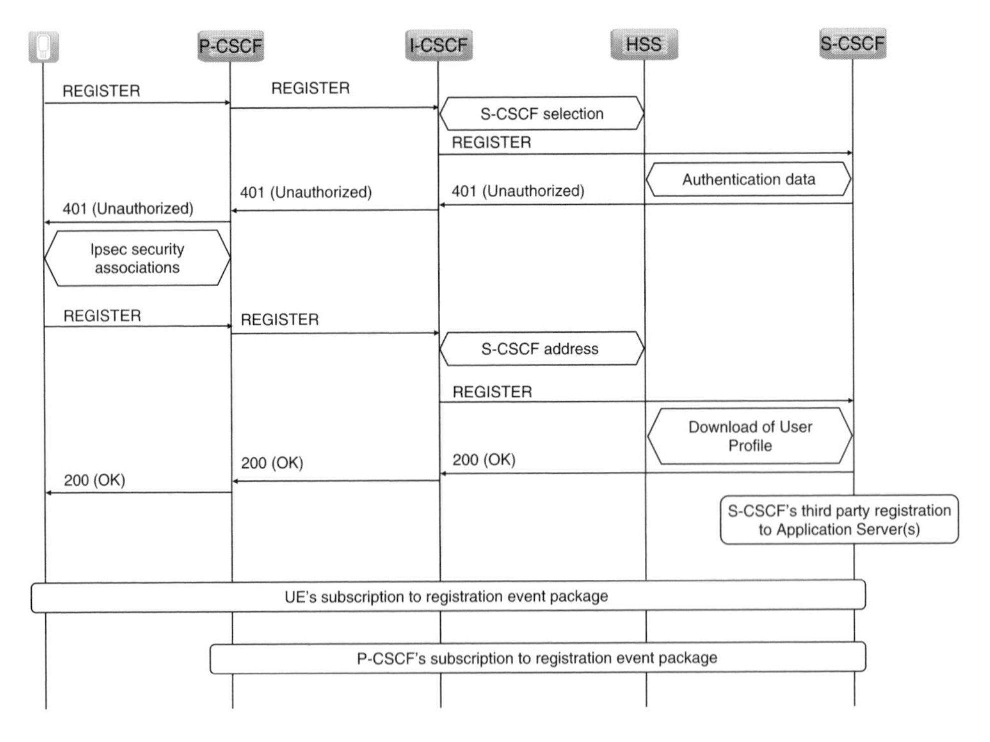

Figure 5.4 IMS registration flow.

5.4.1 *Constructing the REGISTER Request*

After establishing the default bearer and discovering the P-CSCF address, Kristiina's UE can start to construct the initial REGISTER request:

```
REGISTER sip:ims.mnc005.mcc244.3gppnetwork.org SIP/2.0
Via: SIP/2.0/UDP [5555::a:b:c:d]:1400; branch=z9hG4bKnashds7
Max-Forwards: 70
P-Access-Network-Info: 3GPP-E-UTRAN-TDD; utran-cell-id-
     3gpp= 244005F3F5F7
From: <sip:kristiina@example.com>;tag=4fa3
To: <sip:kristiina@example.com >
Contact: <sip:[5555::a:b:c:d]:1400>;
expires=600000;
+sip.instance="<urn:gsma:imei:90420156-025763-0>";
+g.3gpp.smsip;
+g.3gpp.icsi-ref="urn%3Aurn-7%3A3gpp-service-ims.icsi.mmtel"1
Call-ID: apb03a0s09dkjdfglkj49111
Authorization: Digest username="private_user1@example.com",
        realm="ims.mnc005.mcc244.3gppnetwork.org", nonce="",
        uri="sip:ims.mnc005.mcc244.3gppnetwork.org",response=""
Security-Client: ipsec-3gpp; alg=hmac-sha-1-96; spi-c=1111;
     spi-s:=2222; port-c=9999; port-s=1400
Require: sec-agree
Proxy-Require: sec-agree
Supported: path
CSeq: 1 REGISTER
Content-Length: 0
```

The final destination of the request is the SIP registrar (S-CSCF), which is identified in the request URI as page *sip:ims.mnc005.mcc244.3gppnetwork.org* which is the domain name of the home network of Kristiina, as stored in the IP multimedia services identity module (ISIM).[2]

In the To header we find the public user identity sip:kristiina@example.com, which is going to be registered. The From header identifies the user who is performing the registration. We find in the From header the same public user identity as in the To header. Note that the From header includes a tag, while the To header does not. The recipient of the request (i.e. the S-CSCF) will set the To tag when sending the response to the UE. In the Authorization header we find private user identity, private_user1@example.com (as digest username), public user identity, private user identity and home network domain name are obtained from ISIM [or alternatively derived from the universal subscriber identity module (USIM)].

[1] The tag-value-list as defined in (RFC3840) does not support the colon (':') as a valid character. Therefore the colon within the ICSI URNs (urn:urn-7:3gpp-service.ims.icsi.mmtel) needs to be percentage-escaped, as defined in (RFC3986). This means, that every occurrence of the colon character is replaced with the string '%3A' which is the percentage-escaped representation of the colon character.

[2] Here home network domain name obtained from ISIM follows the same format as it would have been derived from USIM.

The Contact header tells that the UE will be reachable for the next 600 000 s under the IP address *5555::a:b:c:d* and port number 1400. This IP address includes the Internet protocol version 6 (IPv6) prefix, which the UE got assigned during establishment of the default bearer (see Section 5.3). The Contact header also contains uniquely device identifier, *urn:gsma:imei:90420156-025763-0*, generated by Kristiina's UE utilising an international mobile equipment identity. In addition, Kristiina's UE signals two feature tags (*+g.3gpp.smsip* and *+g.3gpp.icsi-ref=urn:urn-7:3gpp-service.ims.icsi.mmtel*) in the Contact header. These feature tags reveals that this particular UE supports two specific communication capabilities: SMS over IP (Section 5.9.2.2) and MMTel (Section 4.6).

UE's IP address and port number are also put into the Via header of the request. This ensures that all responses to this request will be routed back to the UE. A branch parameter that uniquely identifies the transaction is also put in the Via header.

A Call-ID header is included, which, together with the value of the CSeq header, identifies the REGISTER transaction.

P-Access-Network-Info indicates to the IMS network which access technology is used to send this request. Here it shows that request is send over LTE and it also contains Cell global ID (CGI) which exposes the location of the user.

In addition the REGISTER request contains additional headers (Security-Client, Require and Proxy-Require) to negotiate a common security mechanism (here only one mechanism is signalled 'ip-sec-3gpp') and to exchange IPSec-related parameters.

Finally, there is the indication that the REGISTER request does not carry any content, as the Content-Length header is set to 0.

5.4.2 From the UE to the P-CSCF

Now Kristiina's UE can send out the REGISTER request to the next hop, which is the discovered P-CSCF (see Section 5.3 and step 11). The request could be sent via the user datagram protocol (UDP) or transmission control protocol (TCP). In order to reach the P-CSCF, the UE puts the following information in the lower layer messages: IP address of P-CSCF and port number as received from domain name system (DNS; or default port 5060 for SIP is used). This request traverses through enodeB and PDN GW to the P-CSCF (see Figure 4.13).

5.4.3 From the P-CSCF to the I-CSCF

When the P-CSCF receives the REGISTER request it sees where it is coming from but as it is not received via IPSec security association (SA) yet, it can only act as a SIP outbound proxy and, therefore, tries to route the REGISTER request to the next hop. The next hop address is found from the request URI as it contains the registrar address. The P-CSCF performs DNS queries with the registrar address (ims.mnc005.mcc244.3gppnetwork.org) to find out an IP address of home network entry point. Prior sending the REGISTER request further the P-CSCF modifies the request as follows: adds itself to the Via header including a branch parameter (in order to receive the response to the request), includes a Path header containing its own address (in order to deliver its own address to S-CSCF for future terminating requests), includes P-Visited-Network-ID header (indicates to the home network the identification of the network where the user is currently located, while roaming this information reveals the IMS roaming partner and it will be used for roaming check), adds an integrity-protected field with the value 'no' to the Authorization header (indicates that the P-CSCF cannot guarantee that the REGISTER request really does originate from the user stated in Authorization header) and a P-Charging-Vector (P-C-V) header containing the IMS charging correlation identifier (ICID) Inter Operator Identifier (IOI; identifying the network which sends the request). After all this the P-CSCF will put the resolved address of home network entry point as the destination address into the UDP/TCP packet that transports SIP requests.

5.4.4 From the I-CSCF to the S-CSCF

The I-CSCF is the entry point to the home network and will receive every REGISTER request. As the I-CSCF has no knowledge about the assignments of user profiles to specific S-CSCFs, it needs to find out to which S-CSCF it should forward the REGISTER request. Therefore it needs to query the HSS (private user identity is used as a key to find correct user information) using the Cx diameter procedures. The HSS performs authorisation checks [e.g. whether the I-CSCF is allowed to query authorisation status of the user and whether the user is allowed to use (roam to) the network as included in P-Visited-Network-ID] and in positive case it will either provide required capabilities for S-CSCF selection or stored address of S-CSCF. If capabilities for S-CSCF were received then the I-CSCF selects the S-CSCF which supports required capabilities. After selecting the S-CSCF, the I-CSCF adds an address of S-CSCF in the Route header and inserts its own entry in the topmost Via header and sends the REGISTER request to the S-CSCF.

5.4.5 S-CSCF Challenges the UE

The S-CSCF realises that the user is not authorised and, therefore, retrieves authentication data from the HSS. Authentication data includes (among other parameters):

- A random challenge (RAND);
- The expected result (XRES);
- The network authentication token (AUTN);
- The integrity key (IK);
- The ciphering key (CK);
- Authentication scheme [IMS authentication and key agreement (AKA)].

In order to authenticate, the S-CSCF rejects the initial REGISTER request from the user with a 401 (Unauthorized) response, which includes (among other parameters) the RAND, the AUTN, the IK, the CK and the authentication scheme.

```
SIP/2.0 401 Unauthorized
WWW-Authenticate: Digest realm="244.91.3gppnetwork.org",
nonce=A34Cm+Fva37UYWpGNB34JP, algorithm=AKAv1-MD5,
ik="0123456789abcdeedcba9876543210",
ck="9876543210abcdeedcba0123456789"
```

The P-CSCF, when receiving the 401 (Unauthorized) response, removes the IK and the CK from the response before sending it to the UE. The IK is the base for the SA that get established between the P-CSCF and the UE immediately afterwards (not described here). This response is sent to UE as shown in Figure 4.13.

5.4.6 UE's Response to the Challenge

The UE, when receiving the 401 response, passes the received parameters over to the ISIM (or USIM) application, which:

- Verifies the AUTN based on the shared secret and the SQN – when AUTN verification is successful the network is authenticated (i.e. the UE can be sure that the authentication data were received from the home communication service provider's network);
- Calculates the result (RES) based on the shared secret and the received RAND;
- Calculates the IK, which is then shared between the P-CSCF and the UE and will serve as the base for the SAs;
- Calculates the CK if signalling ciphering is desired.

With the help of these parameters the UE is able to initiate the second phase of IMS registration, as shown in Figure 5.4. The UE creates the second REGISTER request which contains the same registration related information and will be routed in exactly the same way as the initial REGISTER request. This request has new CSeq numbers, branch parameters, a new From tag in addition to the updated authorisation information and it will be protected by IPSec.

```
REGISTER sip:ims.mnc005.mcc244.3gppnetwork.org SIP/2.0
...
Authorization: Digest username="private_user1@example.com",
realm="ims.mnc005.mcc244.3gppnetwork.org",
nonce=A34Cm+Fva37UYWpGNB34JP, algorithm=AKAv1-MD5,
uri="sip:ims.mnc091.mcc244.3gppnetwork.org ",
response="6629fae49393a05397450978507c4ef1"
```

The P-CSCF, when receiving the second REGISTER request, is now in a position to detect that the received REGISTER request was not modified on its way from the UE to the P-CSCF because it was received via IPSec SA. Therefore, the P-CSCF adds the 'integrity-protected' field with the value 'yes' to the Authorization header and sends the REGISTER request towards the home network.

5.4.7 Registration at the S-CSCF

Eventually the S-CSCF receives the request and compares the received RES from the UE and the XRES that was received from the HSS. If these two parameters are identical, then the S-CSCF has successfully authenticated the user. Once the authentication procedures are successful the S-CSCF will register Kristiina's public user identities. This means the S-CSCF will create a binding for the public user identity that was indicated in the To header of the REGISTER request (sip: kristiina@example.com) and the contact address (sip:[5555::a:b:c:d]:1400). This binding will exist for exactly for 10800 s, which is the value enforced by the S-CSCF based on local policy. This value is considerable shorter value than originally proposed by the UE. The S-CSCF also updates the information in the HSS to indicate that Kristiina has now been registered and it downloads Kristiina's IMS service profile (see Section 5.2) from the HSS.

5.4.8 The 200 (OK) Response

To indicate successful registration the S-CSCF sends 200 (OK) response to the UE.

```
SIP/2.0 200 OK
Via: SIP/2.0/UDP icscf1.example.com;branch=3ictb
Via: SIP/2.0/UDP pcscf1.example.com;branch=2pctb
Via: SIP/2.0/UDP [5555::a:b:c:d]:1400;branch=1uetb
From: <sip:kristiina@example.com>;tag=new
To: <sip:kristiina@example.com>;tag=2222
P-Associated-URI: <sip:kristiina@example.com>,
     <tel:+358501234567>
P-Charging-Function-Addresses: ccf=chargingentity.example.com
P-Charging-Vector: orig-ioi: "Type1  example.com";
     term-ioi: "Type 1 example.com"
Path: <sip:term@pcscf1.home.fi;lr>
Service-Route: <sip:orig@scscf1.home.fi; lr>
Contact: <sip:[5555::a:b:c:d]:1400>;
             Expires=10800;
             +sip.instance="<urn:gsma:imei:90420156-025763-0>";
             +g.3gpp.smsip;
             +g.3gpp.icsi-ref="urn%3Aurn-7%3A3gpp-service-
ims.icsi.mmtel"
Call-ID: apb03a0s09dkjdfglkj49111
CSeq: 2 REGISTER
Content-Length: 0
```

The response is routed back to the UE via all the CSCFs that received the REGISTER request; it manages to do this because CSCFs put their own address in the topmost Via header list when they receive REGISTER requests. Now, when receiving the 200 (OK) response, they just remove their own entry from the Via list and send the request forward to the address indicated in the topmost Via header. The UE, when receiving this response, will know that the registration was successful. The response also contains Path,

Service-Route, P-Associated-URI, P-Charging-Function-Addresses (P-C-F-A) headers and P-C-V headers.The Service-Route header allows to send further requests directly to the S-CSCF without traversing via the I-CSCF. The P-Associated-URI header carries information which other public user identities got registered at the same time (here one additional identity, which could be user's ordinary phone number). The P-C-F-A transports the address of the charging function (here the address of the offline charging function). This enables both S-CSCF and P-CSCF to send charging information to the same network function. P-C-V contains the IOI, a globally unique identifier exposing the sending and receiving networks, here identifying both P-CSCF (orig-ioi as received from P-CSCF in the REGISTER request) and S-CSCF (term-ioi).

5.4.9 Third-Party Registration to Application Servers

After successful registration the S-CSCF will check the downloaded filter criteria of the user (see Section 5.4.7 and Figure 5.5). We assume that there is a telephony application server (TAS) that provides its services to Kristiina; this TAS needs to know that Kristiina has now been registered and is therefore available. To inform the TAS about this, filter criteria have been set which contain a trigger all the REGISTER requests that originate from Kristiina's public user identity. Due to these filter criteria, the S-CSCF will generate a third-party REGISTER request (Figure 5.5) and send it to the TAS whenever Kristiina performs a successful registration.

This REGISTER request is destined to the TAS at TAS.example.com, as indicated in the request URI. As no Route header is included, the request will be sent directly to that address. The To header includes the public user identity of Kristiina, as this is the URI that was registered. The S-CSCF indicates its own address in the From header, as it is registering Kristiina's public user identity on behalf of Kristiina (i.e. as a third

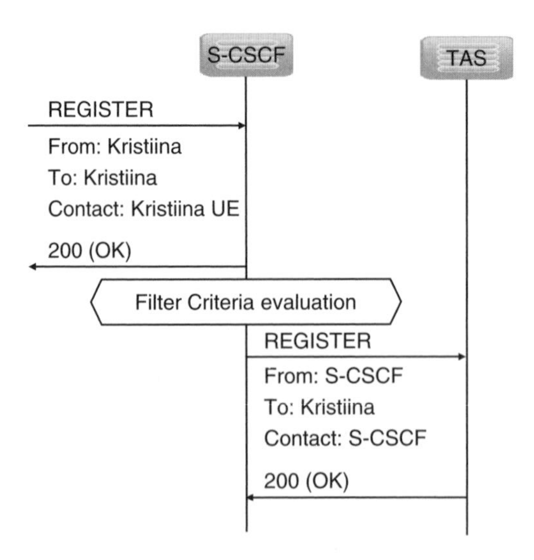

Figure 5.5 IMS third party register by S-CSCF.

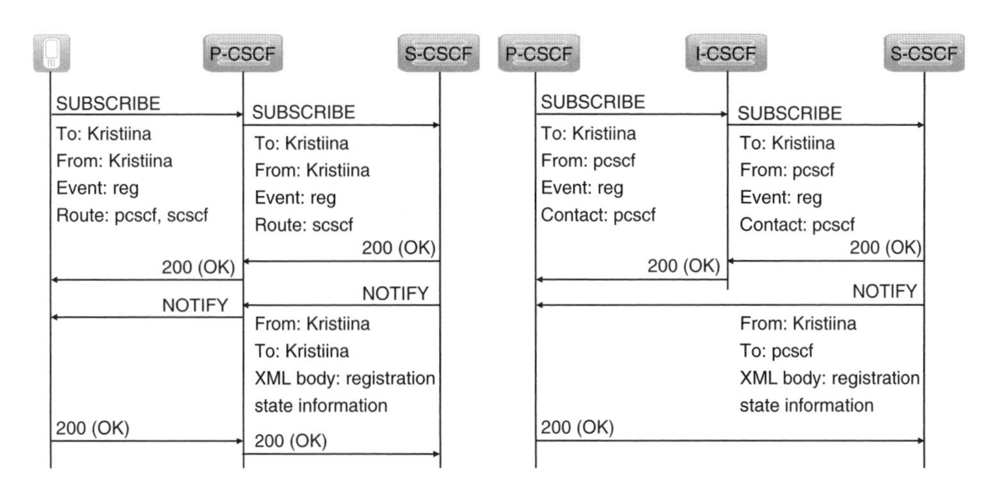

Figure 5.6 Subscription to registration event package.

party). Furthermore, the S-CSCF indicates its own address within the Contact header. This ensures that the TAS never routes directly to Kristiina's UE, but will always contact the S-CSCF first. The S-CSCF may also include the original REGISTER request from the UE inside the body of this third party REGISTER request.

The TAS will send back a 200 (OK) response for this REGISTER request to the S-CSCF, but will not start acting as a registrar for Kristiina. It will take the REGISTER request as an indication that Kristiina has been successfully registered at the S-CSCF. If the TAS needs more information about Kristiina's registration state (e.g. all other public user identities that have been implicitly registered for Kristiina), it can subscribe to the registration-state information of Kristiina in the same way as the UE and the P-CSCF do (see Section 5.4.10).

5.4.10 Subscription to Registration Event Package

After the registration and authentication has succeeded, both UE and P-CSCF need to subscribe to the registration state event package (Figure 5.6). This is needed in order to receive network initiated re-authentication (see Section 5.4.11) and receive information regarding network initiated de-registration (see Section 5.4.12). In addition, registration state event package carries registration status about all available public user identities (all information related to subscription not limited to this particular UE). This status information reveals, for example if particular public user identity is also registered from other UE.

5.4.11 Re-Registration and Re-Authentication

The IMS UE registers its contact information for a time of 10800 s, which means that the binding of the registered public user identities and the physical IP address is kept for around three hours in the S-CSCF. It is the UE's responsibility to keep its registration active by periodically refreshing its registration, that is before the registration timer

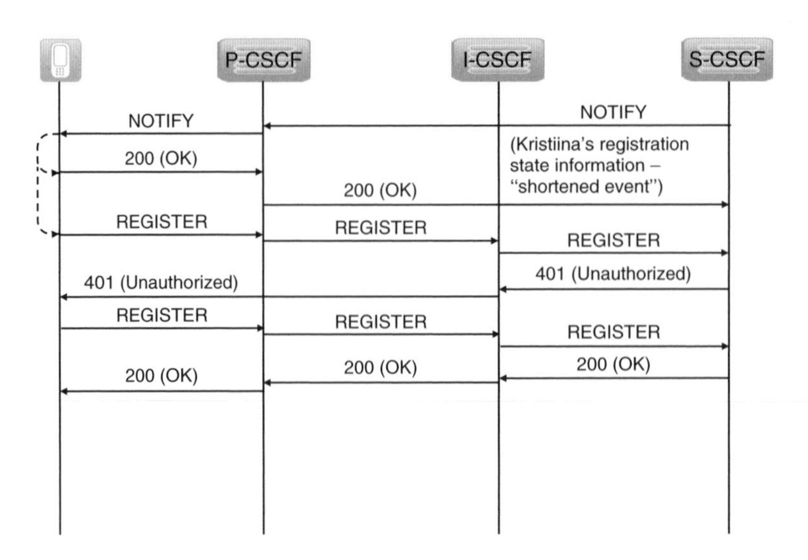

Figure 5.7 Network initiated re-authentication.

lapses in the S-CSCF. To maintain a registration active the UE sends a new REGISTER request to the network. This re-registration is handled in the same way as an initial SIP registration procedure in the network as described in previous sections. If the UE does not refresh its registration, then the S-CSCF will silently remove the registration when the registration timer lapses.

User authentication procedures are directly coupled to registration procedures. Therefore the S-CSCF needs to be able to shorten the registration time, in order to re-authenticate the UE (Figure 5.7). To achieve this, the S-CSCF reduces the expiry time of the user's registration, for example to 300 s. The new expiry time is delivered to the UE utilising the UE's subscription to the registration-state event package. In other words the S-CSCF generates a NOTIFY request for the registration-state event package, in which it indicates that it shortened the registration time and sends this NOTIFY request to the registered UE. On receiving this request the UE will immediately update the registration expiry time information. Furthermore, all other subscribers to the user's registration-state information (e.g. the P-CSCF and the subscribed ASs) will receive a NOTIFY request from the S-CSCF with the updated state information. To maintain active registration the UE will send out a REGISTER request. From then on, the normal registration procedures will take place, during which the S-CSCF can authenticate the user again (see Sections 5.4.5 and 5.4.7).

5.4.12 De-Registration

All things come to an end at some point, and this is also true for the registration of a user to the IMS. Kristiina might want to be undisturbed and switches off her mobile phone. When doing so, her phone sends another REGISTER request to the S-CSCF, including all the information we have already seen, but indicating that this time it is for de-registration by setting the expiry time to zero (Figure 5.8). The S-CSCF will then clear all the temporary information it has stored for Kristiina, update the data in the HSS and send a 200 (OK) response to Kristiina's UE.

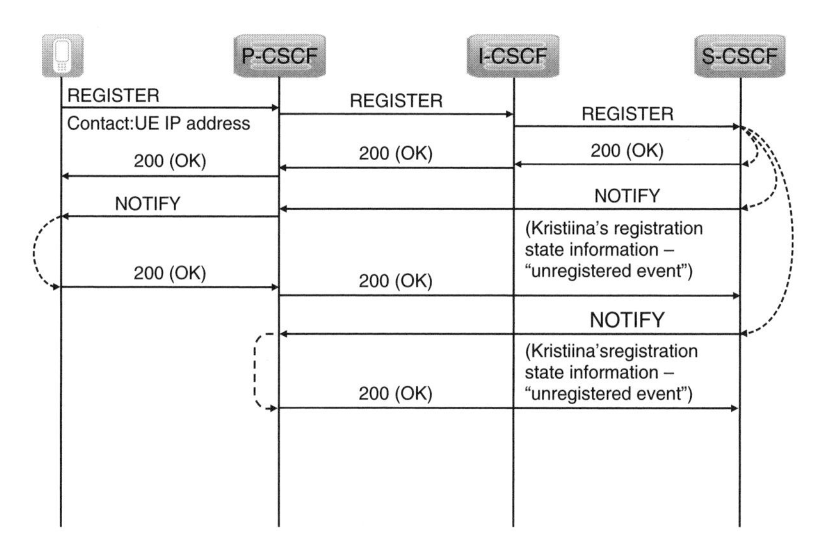

Figure 5.8 UE initiated de-registration.

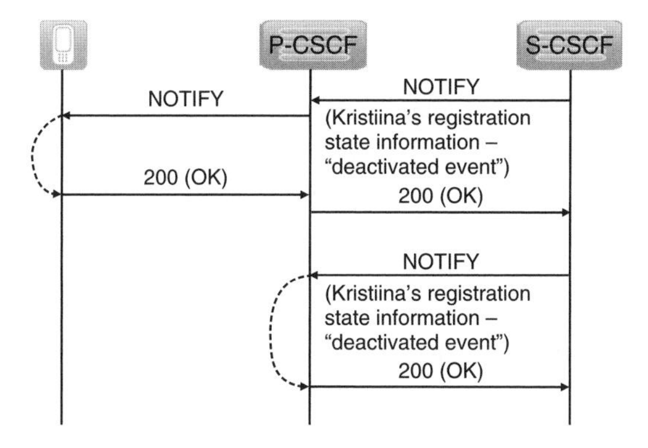

Figure 5.9 Network initiated de-registration.

Sometimes the network needs to de-register the user (Figure 5.9), for example when UE is reported stolen, unpaid bills, S-CSCF needs to be shut down for administrative reasons. In these cases the S-CSCF would simply send a NOTIFY message with registration-state information to Kristiina's UE, this time indicating that she has been de-registered.

5.4.13 Related Standards

Specifications relevant to Section 5.4 are:

- 3GPP TS 23.003 – Numbering, addressing and identification.
- 3GPP TS 23.228 – IP multimedia subsystem (IMS).
- 3GPP TS 24.229 – IP Multimedia call control protocol based on SIP and SDP.
- 3GPP TS 29.228 – IMS Cx and Dx interfaces – signalling flows and messages.
- 3GPP TS 29.229 – IMS Cx and Dx interfaces based on the diameter protocol.

- 3GPP TS 33.102 – Security architecture.
- 3GPP TS 33.203 – Access security for IP-based services.
- RFC2401 – Security architecture for the Internet protocol.
- RFC2617 – HTTP Authentication: basic and digest access authentication.
- RFC3261 – SIP: session initiation protocol.
- RFC3265 – Session initiation protocol (SIP)-specific event notification.
- RFC3310 – Hypertext transfer protocol (HTTP) digest authentication using authentication and key agreement (AKA).
- RFC3325 – Private extensions to the session initiation protocol (SIP) for asserted identity within trusted networks.
- RFC3327 – Session initiation protocol (SIP) extension header field for registering non-adjacent contacts.
- RFC3329 – Security mechanism agreement for the session initiation protocol (SIP).
- RFC3455 – Private header (P-header) extensions to the session initiation protocol (SIP) for the 3rd generation partnership project (3GPP).
- RRC3608 – Session initiation protocol (SIP) extension header field for service route discovery during registration.
- RFC3680 – A session initiation protocol (SIP) event package for registrations.
- RFC4566 – SDP: session description protocol.
- RFC5279 – A uniform resource name (URN) namespace for the 3rd generation partnership project (3GPP).
- http://www.3gpp.org/tb/Other/URN/URN.htm – URN values maintained by 3GPP.
- RFC6050 – A session initiation protocol (SIP) extension for the identification of services.

5.5 IMS VoIP Session

This section shows how our example users Kristiina and Jennifer establish a voice session (see Figure 5.10) and what happens during this procedure. The SIP messages between UE and P-CSCF are traversing as described in Section 4.3.2 and depicted in Figure 4.13.

The first step in the session established is to create a session offer based on the user's desire and UE capabilities. Section 5.5.1 shows a global system for mobile communication association (GSMA) (IR.92) compliant voice session offer. After the creation of an initial request we need to find a way to the desired recipient. This we call here as routing and this is in fact one of the most complex issues within IMS. This and how response and subsequent request are routed are covered in Section 5.5.2. During the session setup phase the two UEs agree on the set of media they want to use for the session and the codecs which will be used for the different media types. This procedure is covered in Section 5.5.3. In addition, we need to secure that the required user plane resources to carry actual voice packets are in place on both sides before Jennifer's UE is allowed to alert her device regarding the incoming voice session (see Section 5.5.4). In this example we do not show the detailed interaction with the ASs (see Section 4.5.5 how originating and terminating sessions are routed to AS). For supplementary service execution see Section 4.6, and for SR-VCC procedures see Section 5.6.2. This section is concluded with subsections which describe how the charging of a voice session (Section 5.5.5) is done and how an established session is released (Section 5.5.6).

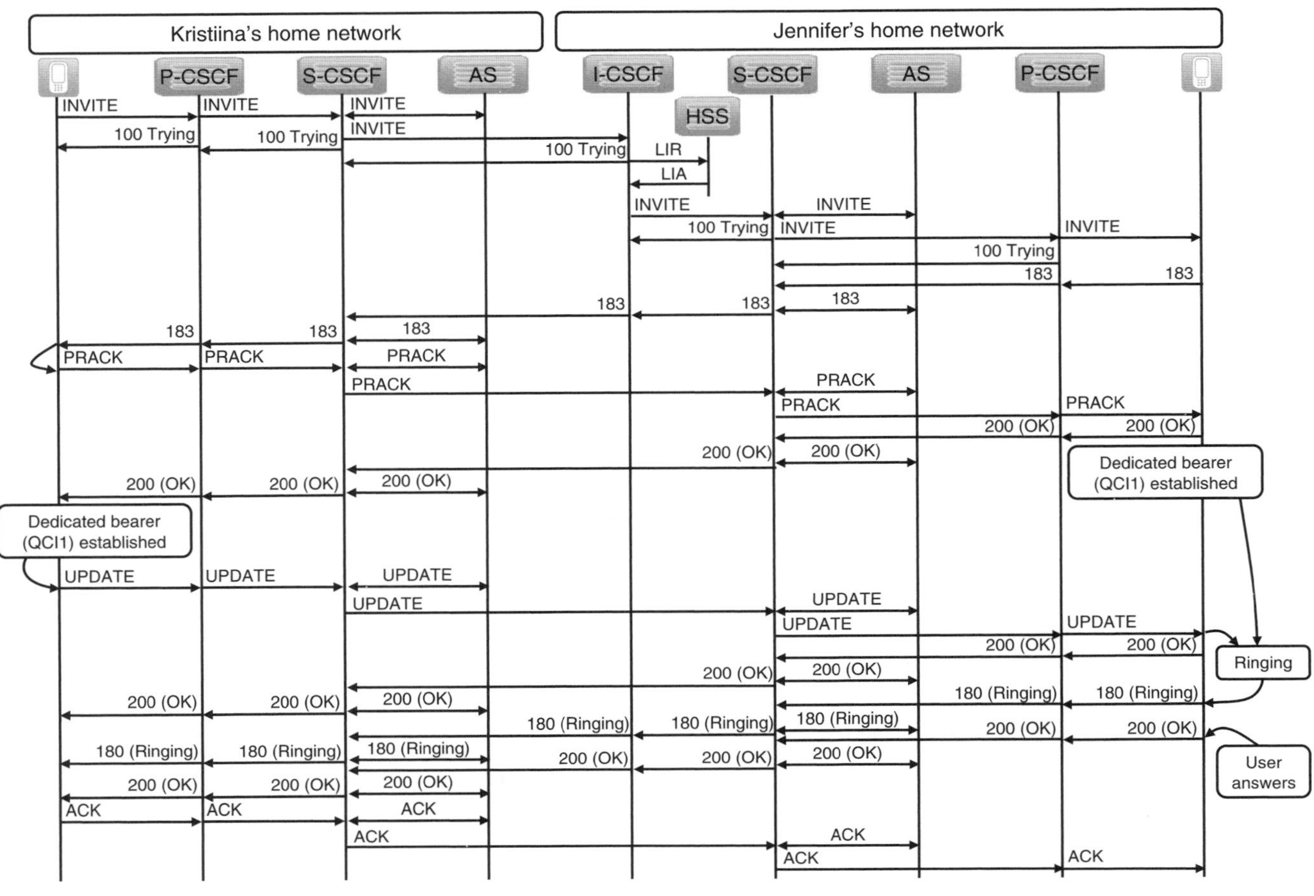

Figure 5.10 IMS session initiation.

5.5.1 Constructing the INVITE Request

For establishing a voice communication via the IMS, UEs uses the SIP and session description protocol (SDP) to setup the session and to agree communication media components such as audio.

SIP

```
INVITE sip: jennifer@csp.com SIP/2.0
Via: SIP/2.0/UDP [5555::a:b:c:d]:1400; branch=abc123
Max-Forwards:70
Route: <sip:[5555::55:66:77:88]:7531;lr>,< sip:orig@scscf1.home.fi;lr>
P-Access-Network-Info:3GPP-E-UTRAN-TDD;utran-cell-id-3gpp=244005F3F5F7
P-Preferred-Service: urn:urn-7:3gpp-service.ims.icsi.mmtel
Privacy: none
From: <sip:kristiina@example.com>;tag=171828
To: <sip:jennifer@csp.com>
Call-ID: cb03a0s09a2sdfglkj490333
Cseq: 127 INVITE
Require: sec-agree
Proxy-Require: sec-agree
Supported: precondition, 100rel, 199
Security-Verify: ipsec-3gpp; alg=hmac-sha-1-96; spi-c=98765432;
spi-s=87654321; port-c=8642; port-s=7531
Contact: <sip:[5555::a:b:c:d]:1400;+g.3gpp.icsi-ref="urn%3Aurn-7%
3gpp-service.ims.icsi.mmtel"
Accept-Contact: *;+g.3gpp.icsi-ref="urn%3Aurn-7%
3gpp-service.ims.icsi.mmtel"
Allow: INVITE, ACK, CANCEL, BYE, PRACK, UPDATE, REFER, MESSAGE, OPTIONS
Accept:application/sdp, application/3gpp-ims+xml
Content-Type: application/sdp
Content-Length: (...)
```

SDP

```
v=0
o=- 2890844526 2890842807 IN IP6 5555::a:b:c:d
s=-
c=IN IP6 5555::a:b:c:d
t=0 0
m=audio 49152 RTP/AVP 97 98
a=rtpmap:97 AMR/8000/1
a=fmtp:97 mode-change-capability=2; max-red=220
b=AS:30
b=RS:0
b=RR:0
a=rtpmap:98 telephone-event/8000/1
a=fmtp:98 0-15
a=ptime:20
a=maxptime:240
a=inactive
a=curr:qos local none
a=curr:qos remote none
a=des:qos mandatory local sendrecv
a=des:qos none remote sendrecv
```

Let us have a closer look at this.

5.5.1.1 SIP Information

- The first line tells us the final destination of the INVITE request – which is the SIP URI of Jennifer. The purpose of the Via, Route and Contact headers is covered in Section 5.5.2.
- The Max-Forwards header contains an integer that is decremented by one at each hop to detect a loop (initially it is set to 70).
- The P-Access-Network-Info header is a 3GPP-specific header and indicates the IMS network over which access technology (LTE) the UE is attached to IMS. It also includes the CGI, which indicates the location of the user.
- P-Preferred-Service header is a 3GPP-specific header and indicates the IMS communication service (here MMTel) that is related to the request.
- The Privacy header is used to signal whether the user wants to hide or make her identity available for other users (see Sections 4.6.3.1–4.6.3.4).
- From and To header values do not play a role on IMS routing as in the IMS end-user trusted identifiers are signalled with P-Asserted-Identity header (not visible here as this header is provided by the network not the UE), see for example Section 4.6.3.1.
- Every SIP dialogue is identified by the value of the Call-Id header and by tags in the To and the From headers of the SIP requests (no tag present in To header it is set by Jennifer's UE).
- The CSeq header field serves as a way to identify and order transactions (here with a value of 127). Every subsequent request sent from the same end (here Kristiina's UE) will have a higher CSeq than the preceding request (e.g. PRACK request includes 128, UPDATE 129 and so forth).
- The Require, Proxy-Require and Security-Verify headers are needed to ensure that IPSec is applied as agreed during IMS registration.
- The Supported header lists the SIP extensions which the Kristiina's UE is able to use. Here three important extensions (precondition, 100rel, 199) are supported as mandatorily required in 3GPP and GSMA (IR.92) specifications. '100rel' extension means that Kristiina's UE will explicitly acknowledge when it receives 100-class response (e.g. 183 Session Progress) from Jennifer's UE. Without this extension, Jennifer's UE that sends 183 Session Progress containing the final codec does not get any indication back whether this response was ever received by the Kristiina's UE. The '199' extension means that Kristiina's UE supports SIP response code, 199 'Early Dialog Terminated', which is used to signal that an early SIP dialogue has been terminated before a final response is sent towards the Kristiina's UE. Preconditions extension is further covered in Section 5.5.4.
- The Accept-Contact header expresses a preference of the caller (Kristiina) to reach a UE of the called party (Jennifer), which support the indicated capabilities [here IMS multimedia telephony communication service (MMTel), see also Section 4.6 for more detailed description of MMTel service]. The called UE also adds all its own supported capabilities within the Contact header.
- Finally the SIP contains the Allow header, which lists the set of methods supported by the user agent (UA) generating the message. The Accept header lists that Kristiina's UE supports SDP and IMS special extensible markup language (XML) payload, and the Content-Length header indicates the size of the message-body (here SDP).

5.5.1.2 SDP Information

- The v-line indicates the protocol version and is always set to 0.
- The o-line holds parameters related to the owner of the session, who is in this case Kristiina.
- The s-line may include a subject for the session; here empty.
- The c-line contains information about the connection that has to be established for the multimedia session: that is this indicates the addresses used for the real media streams.
- The t-line, which indicates when the session was created and how long it is intended to last. There need be no time limitation to the session set in SDP, as SIP users already end a session by manually sending a BYE request. So, the parameters of the t-line can safely be given as 0.
- The rest of the SDP parameters are described in Sections 5.5.3 and 5.5.4.

5.5.2 Routing

Kristiina's UE is unaware at the time of sending the INVITE request how Jennifer's UE can be reached. All it can provide is:

- The final destination of the INVITE request – this is the SIP URI of Jennifer (one of her public user identities, jennifer@csp.com) that Kristiina had to indicate (e.g. by selecting it from her phone book).
- The address of the P-CSCF (sip:[5555::55:66:77:88]:7531) – this is the outbound proxy of Kristiina's UE and will be the first hop to route to. This address is obtained before SIP registration during the P-CSCF discovery procedures (see Section 5.3 and step 11).
- The address of the S-CSCF (sip:orig@scscf1.home.fi) – this was discovered during registration procedures by means of the Service-Route header (see Section 5.4.8).

Armed with this partial route information the INVITE request is sent on its way. It first traverses the P-CSCF and then the S-CSCF that have been selected for Kristiina. The S-CSCF executes services as described in Section 4.5.5 and after that the S-CSCF has no further routing information available for the request other than the final destination (i.e. the public user identity of Jennifer, sip: jennifer@csp.com). As Kristiina's S-CSCF does not act as a registrar for Jennifer, it can only resolve the host part of the address: csp.com. This domain name is sent to the DNS server and the S-CSCF will receive back one or more I-CSCF addresses of Jennifer's home network, will select one of them and will send the INVITE request to it.

The I-CSCF just acts as the entry point to Jennifer's home network. Figure 5.10 shows that the I-CSCF asks the local HSS for the address of the S-CSCF that was selected for Jennifer with a Location Information Request (LIR). The HSS returns the address of S-CSCF in a Location Information Answer (LIA). The I-CSCF sends the INVITE further on to the S-CSCF. Jennifer's S-CSCF executes services (see Section 4.5.5) and after that it acts as the registrar and replaces Jennifer's SIP URI in the Request URI of the INVITE request with the contact address that she has registered. It does not send the request directly to Jennifer's UE, because it remembers the Path header information from registration. The INVITE request, therefore, is first sent to Jennifer's P-CSCF. The S-CSCF knows the address of the P-CSCF, as that was received within the Path header during Jennifer's registration (see Section 5.4.3). The P-CSCF finally forwards the INVITE request to Jennifer's UE over the IPsec SA. This shows that for the initial request the route from Kristiina to Jennifer is put together piece by piece, as the originating UE and the CSCFs have only information about the next one or two hops that have to be traversed. In order to make further routing within the dialogue easier, SIP routing mechanisms will be used:

- All CSCFs put their addresses on top of the Via header – this allows all responses to the INVITE request to be sent back over exactly the same route as the request.[3]
- All CSCFs, other than Jennifer's I-CSCF, put their addresses on top of the Record-Route header – this allows all subsequent requests in the dialogue to be sent over the CSCFs that put themselves in the Record-Route header. The I-CSCF in Jennifer's home network fulfilled its routing task when it discovered the addresses of Jennifer's S-CSCF; so, it is no longer needed on the route.

Jennifer's UE now creates a response to the received INVITE request which is, due to the usage of preconditions (see Section 5.5.4), a 183 (Session Progress) response. The UE puts its own IP address in the Contact header to indicate the address it wants to use to receive subsequent requests in this dialogue. The contact address also includes the protected server port of Jennifer's UE (1006), which guarantees that all subsequent requests will be received via the established IPsec SA as well. The Record-Route and Via headers of the INVITE request also go into the response. After doing so, Jennifer's UE sends the response to the address and port number of the topmost entry in the Via header, which is the protected server port of the P-CSCF. The P-CSCF identifies that this is a response to the INVITE and it removes its own address from the Via header and sends the response to the topmost entry in the Via header. Similarly other CSCFs and ASs removes their own Via entry and send the message towards the next entry in the Via.[4] After receiving the response, Kristiina's UE:

- Stores the IP address of Theresa's UE, as received in the Contact header;
- Stores the Record-Route list after reversing the order of all entries in it.

Figures 5.11 and 5.12 show our example in action (for simplicity HSS and ASs are not shown).

When one of the two UEs needs to send a subsequent request within a dialogue, it copies the stored Record-Route entries into the Route header of the new requests and the remote UE's IP address into the Request URI. The request then is routed towards the remote UE by strictly following the entries in the Route header (Figure 5.12). Every CSCF that is traversed puts itself in the Via header, in order to get all the responses to this request. As the I-CSCF did not record any route in the beginning, it does not receive

[3] Likewise TASs (see Section 4.6.3) and Service Centralisation and Continuity AS (see Section 5.6.2) will include their addresses on the Via header and Record-Route (not shown here for simplicity).

[4] CSCFs may re-write their Record-Route entries to distinguish requests received from different directions. P-CSCFs are expected to re-write Record-Route due to use of IPSec (IPSec secured port number is only present between UE and P-CSCF). This type of detail is not shown in Figures 5.11 and 5.12.

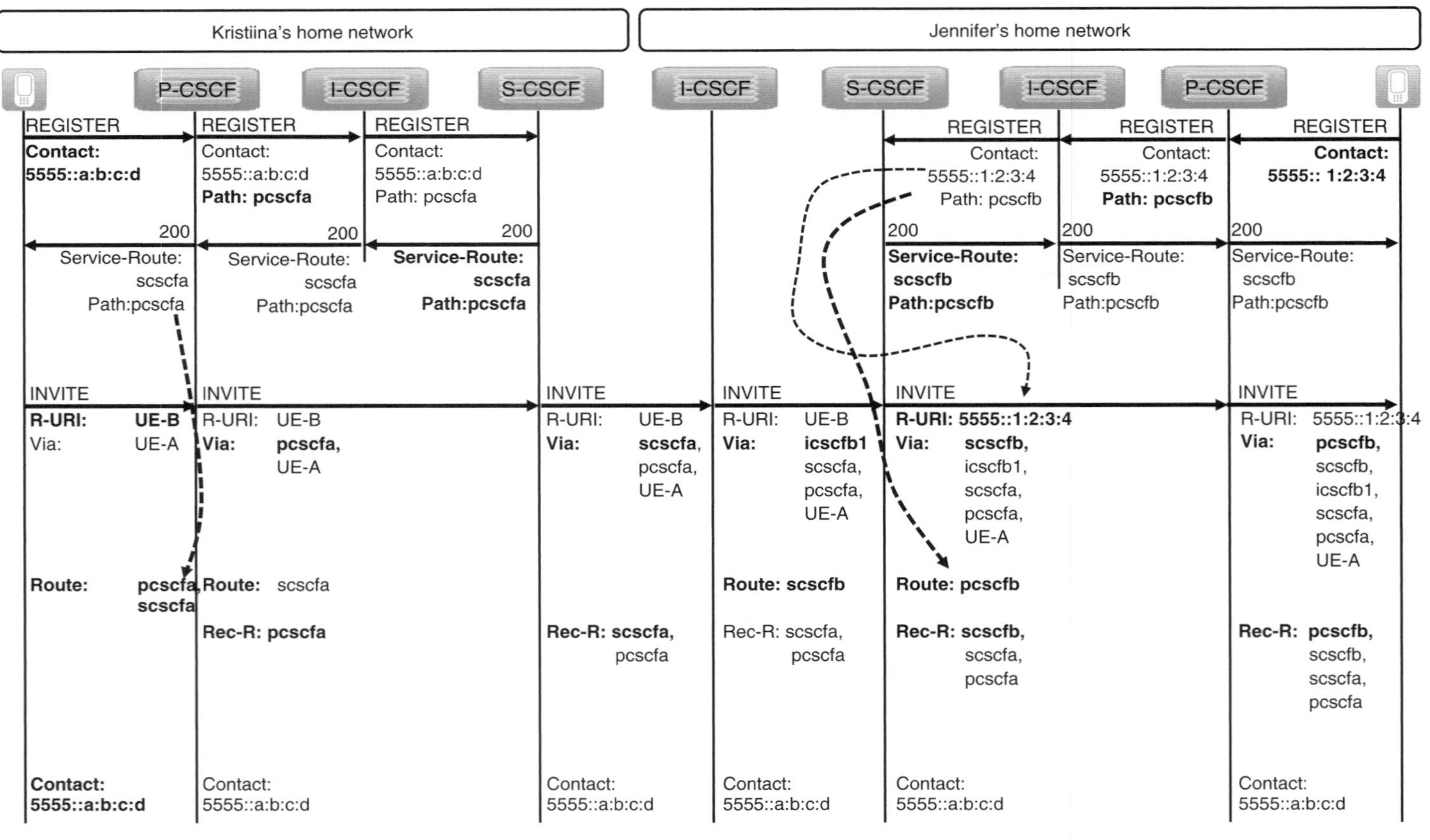

Figure 5.11 Routing an initial INVITE request.

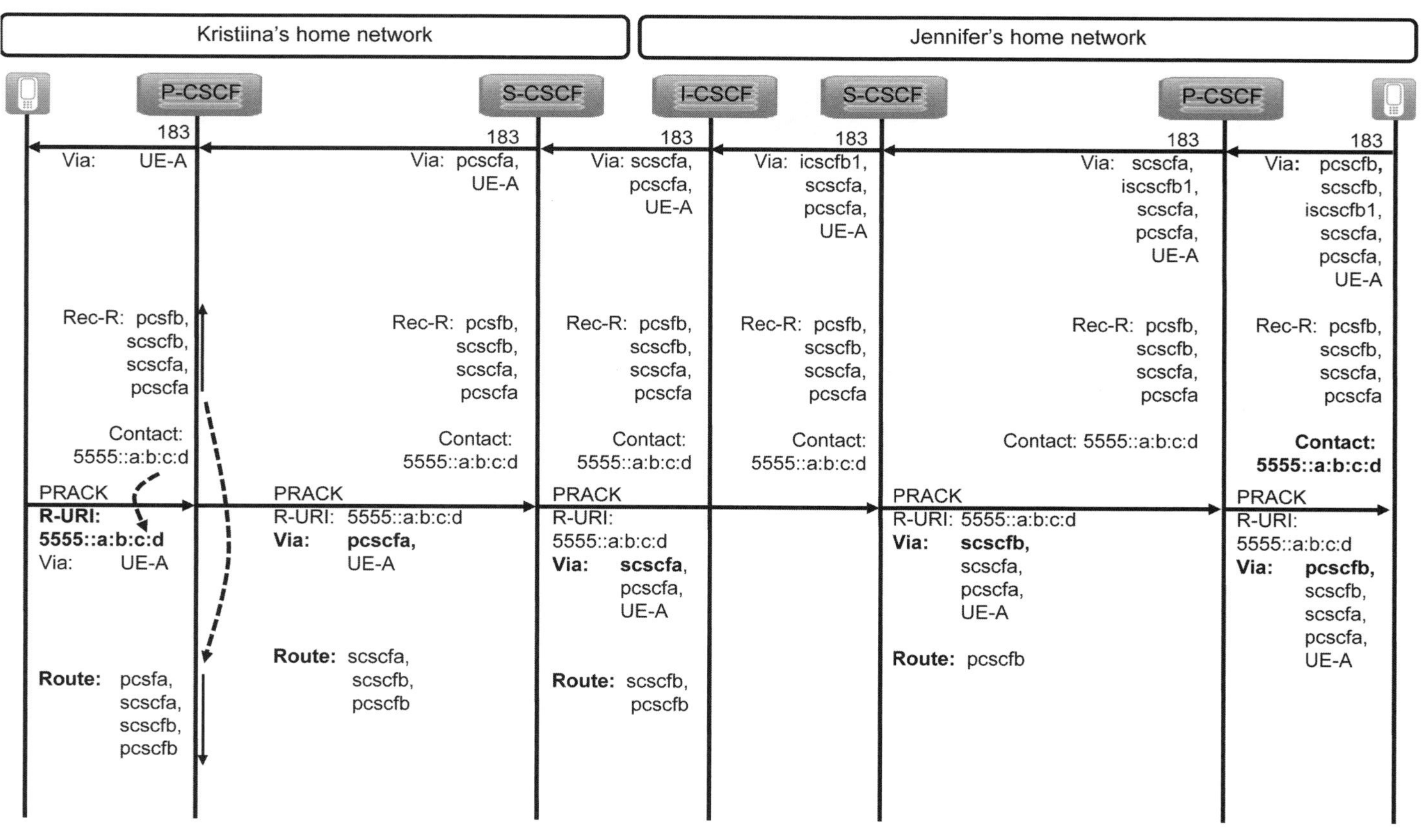

Figure 5.12 Routing a response to the initial INVITE request and subsequent request.

any subsequent request. For example, Tobias's UE has to send back a PRACK request to acknowledge the received 183 (Session Progress) response. (see Section 5.2).

A subsequent request within a dialogue does not include a Contact header, as the addresses of the two UEs were already exchanged during the sending and receiving of the initial request and its first response. Furthermore, the CSCFs will not put any Record-Route headers in the request, because the route was already recorded during the initial request. Theresa's UE will send back a 200 (OK) response to this PRACK request. This response will be routed back on the basic of the Via header entries. Record-Route headers are no longer returned.

Figures 5.13 and 5.14 summarise the IMS routing principles.

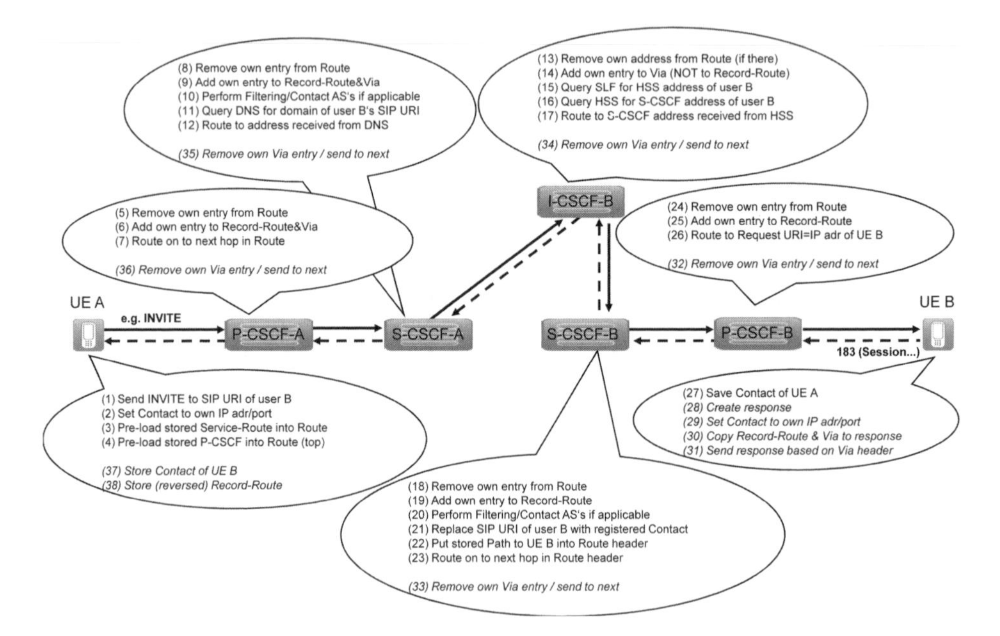

Figure 5.13 Routing principles of initial request and response to it.

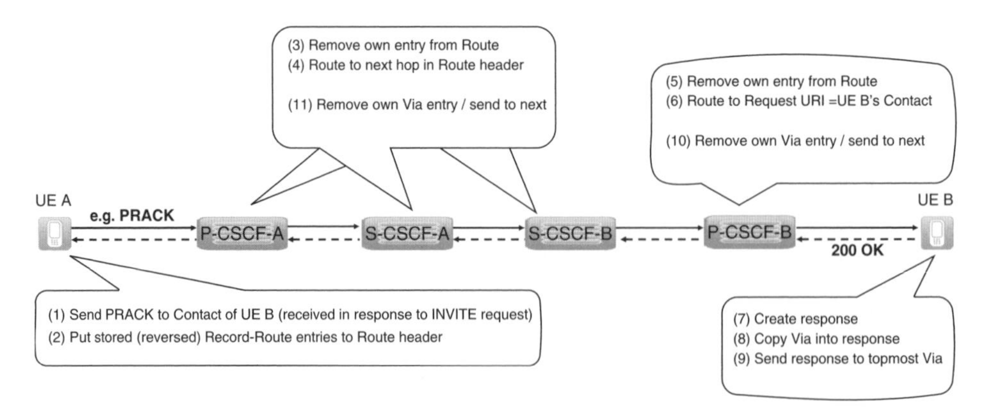

Figure 5.14 Routing principles of subsequent requests and their responses.

5.5.3 Media Negotiation

Section 5.5.1 contained a full-blown example of INVITE request. Here we will take closer look at SDP of it and how it is used agreed media types and the codecs for a specific session.

Here we have modified our earlier INVITE example such way that we have added a second audio codec and video media to the initial SDP in order to demonstrate IMS media negotiation capability. In addition, parameters which are not relevant for the media and codec negotiation procedure point of view are ignored.

```
m=audio 49152 RTP/AVP 97 98 99
a=rtpmap:97 AMR/8000/1
a=rtpmap:98 G.726-32/8000
a=rtpmap:99 telephone-event/8000/1
m=video 3400 RTP/AVP 98
a=rtpmap:98 H264/90000
```

Individual media lines, or m-lines, represent the two different media streams that the user wants to send:

```
m=audio 49152 RTP/AVP 97 98 99
m=video 3400 RTP/AVP 98
```

The first line indicates that Kristiina intends to use audio for this session and that her UE will use the local port 49152. Real-time transport protocol (RTP)/audio video profile (AVP) will be used as the transport protocol for audio-related media. Different dynamic payload types are indicated with the numbers 97–99. These three numbers are further mapped in adaptive multi-rate (AMR) (a=rtpmap:97), G.726 (a=rtpmap:98) speech codecs and telephone event[5] representation of dual-tone multi-frequency (DTMF) tones (a=rtpmap:99). Similarly, the second m-line indicates that Kristiina intends to use video for this session and that her UE will use the local port 3400 and payload type is mapped to H.264 codec.

This SDP offer is sent within the body of the INVITE request to Jennifer as described in Section 5.5.2. The SDP offer arrives at Jennifer's UE due to SIP routing of the INVITE request (as shown in Figure 5.10).

Jennifer's UE generates an SDP answer for the received offer, which looks like this:

```
m=audio 49540 RTP/AVP 98 99
a=rtpmap:98 AMR/8000/1
a=rtpmap:99 telephone-event/8000/1
m=video 0 RTP/AVP 98
```

From the audio m-line we see that only two payload types are left. In IMS the terminating UE, here Jennifer's UE, is responsible to select a single codec per media and

[5] The telephone event defines a text-based representation of these DTMF tones and other telephone-related tones that can be transported over RTP.

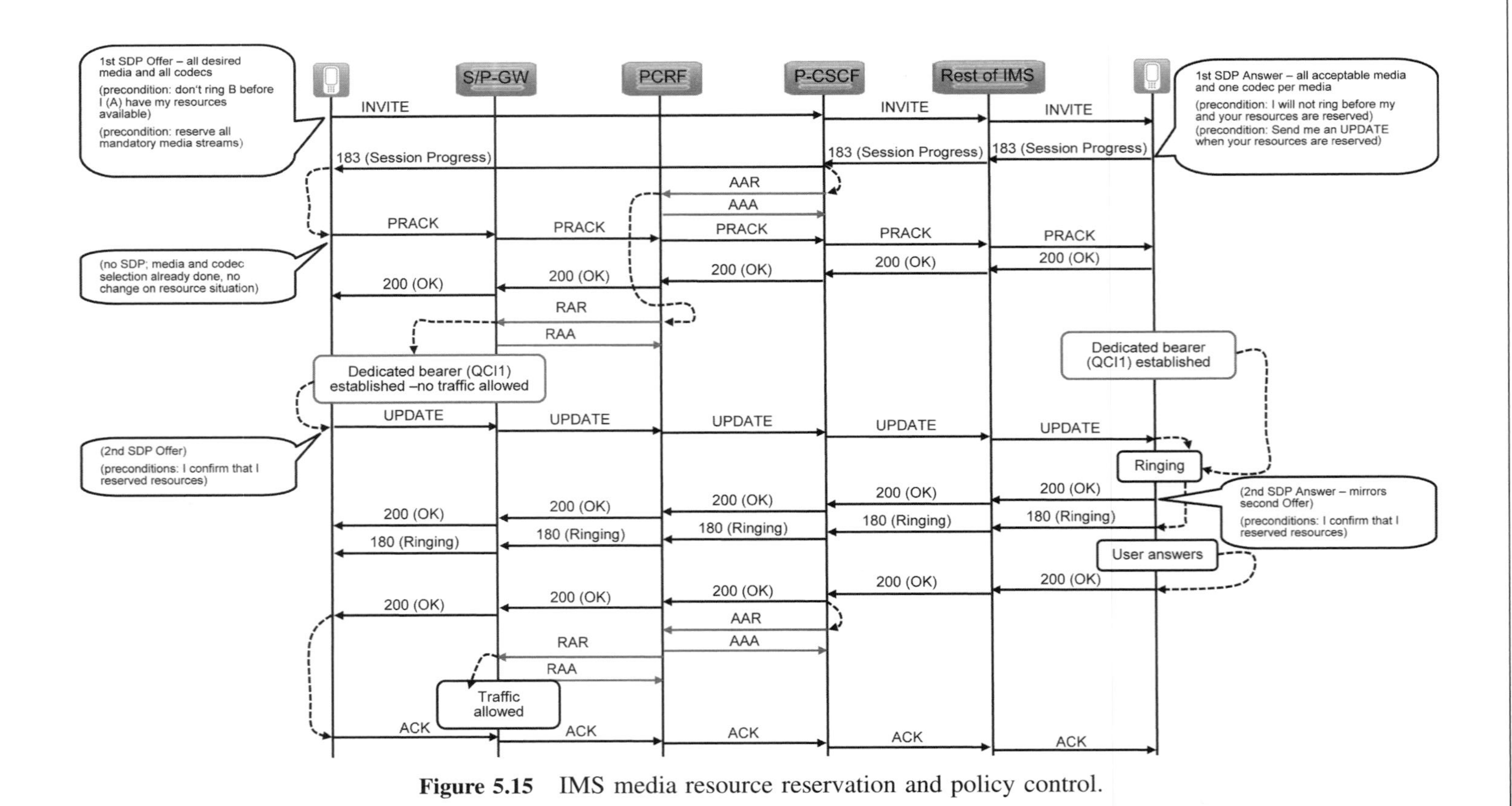

Figure 5.15 IMS media resource reservation and policy control.

here the AMR-NB speech codec was selected and another codec was simply dropped.[6] In addition Jennifer's UE was not able to support video therefore the port number in the second m-line is set to zero and related a-lines are simply dropped. The SDP answer must repeat all the m-lines that are included in the SDP offer. Therefore, the UE cannot drop the second m-line from here: the only way to indicate that this video cannot be handled is to set the port number to 0. This SDP answer is contained in the 183 (Session Progress) response, as shown in Figure 5.15.

5.5.4 Media Resource Reservation and Policy Control

Section 5.5.1 contained a full-blown example of the INVITE request. Here we will take closer look at the SDP of it and how it is used to signal status of resources for session media traffic. In our example both UEs are LTE connected and make use of dedicated bearers for voice packets (see Section 4.3.3). These dedicated bearers need to be established during the IMS session setup. The establishment of dedicated bearers may consume some time and may even fail: when, say, insufficient resources are available over the wireless link. This means that, until the bearers have been established, it cannot be guaranteed that the agreed media connection can be established at all. Therefore, Jennifer's UE should not inform her about the incoming session request (INVITE): that is it should not start to ring until it has confirmation that the required resources have been successfully reserved locally as well as at the calling user's end. In order to achieve this, both UEs utilise SDP precondition mechanism introduced in (RFC3312) that allows a UE to delay completion of a SIP session establishment until one or both ends have successfully completed their resource reservation. This extension to SIP and SDP is mandatorily supported by every UE that connects to IMS. Figure 5.15 provides an overview of media resource reservation and policy control.

Section 5.5.1 contained a full-blown example of INVITE request. In order to illustrate the media resource mechanism, we take just focus on SDP parameters which are required to setup an appropriate dedicated bearer and to ensure that Jennifer is not alerted prior IMS media resources are in place.

```
m=audio 49152 RTP/AVP 97 98
a=rtpmap:97 AMR/8000/1
a=rtpmap:98 telephone-event/8000/1
b=AS:30
b=RS:0
b=RR:0
a=inactive
a=curr:qos local none
a=curr:qos remote none
a=des:qos mandatory local sendrecv
a=des:qos none remote sendrecv
```

[6] If the terminating UE does not select single codec per media then the originating IMS UE needs to take final codec decision and signal the selected codec in the next request, PRACK.

First, we look at the penultimate line:

```
a=des:qos mandatory local sendrecv
```

This indicates the desired (des) QoS precondition at the calling user (local) end. The resources for the calling user need to be reserved in both the sending and receiving directions (sendrecv), as the audio stream is bidirectional (i.e. both users can talk to each other and hear what the other says). It also states that the session will not take place if the indicated resources cannot be reserved by the calling user (mandatory). The last line:

```
a=des:qos none remote sendrecv
```

indicates the desired (des) QoS-related preconditions at the called user (remote) end. As the calling and the called UE are not directly connected to each other, the calling UE is unaware of how the remote UE is attached to the network. It might be that Jennifer's UE does not need to perform any resource reservation, as it is connected via a CS telephony network. Therefore, the calling end can only indicate that, if the remote end needs to reserve resources, then they should be reserved in both the sending and receiving directions (sendrecv); however, the calling end is currently unaware that this is really required in order to get a media session established (none).

Up to now, the calling UE has only expressed the desired (des) preconditions for each end, but it also needs to talk about the current status of resource reservation. This is the subject of the following two new lines:

```
a=curr:qos local none
a=curr:qos remote none
```

These two lines indicate that currently (curr) no (none) QoS-related preconditions have been fulfilled by either the calling end (local) or the called end (remote). The a=des lines make it possible to set preconditions for the local and the remote user, while the a=curr lines indicate the extent to which the set preconditions are already being fulfilled.

In addition to precondition related parameters the SDP contains information about the desired media in the first three lines, as explained in Section 5.5.3. The next three lines:

```
b=AS:30
b=RS:0
b=RR:0
```

reveal that UE has calculated that required bandwidth for RTP traffic is 30 kbit/s and it expresses that it is not going to use RTP control protocol (RTCP; b:RS and b:RR values set to zero) while engaged on this call.[7]

[7] GSMA (IR.92) states that UEs engaged on active call are mandated to disable RTCP as the policy control functionality takes care of supervising user plane link aliveness. Using RTCP for voice media would not bring additional benefit but it would consume upto 5% more bandwidth. Use of RTCP is required when a session is put on hold.

The seventh lines tells us that Kristiina's UE wants to make sure that the other end does not send user plane traffic prior the resources are ready. Later on when the network has established a dedicated bearer then Kristiina's UE will send media status to allow traffic in both directions.

```
a=inactive
```

Here we have assumed that Jennifer's UE is also attached to LTE and supports the preconditions mechanism. Therefore, it will respond to the received SDP offer with a well-formed answer and will include its own preconditions. Once again, we only show here those preconditions and key media-related information:

```
m=audio 49540 RTP/AVP 98 99
a=rtpmap:98 AMR/8000/1
a=rtpmap:99 telephone-event/8000/1
b=AS:30
b=RS:0
b=RR:0
a=inactive
a=curr:qos local none
a=curr:qos remote none
a=des:qos mandatory local sendrecv
a=des:qos mandatory remote sendrecv
a=conf:qos remote sendrecv
```

The important thing to note here is that the remote and local ends have changed, because from Jennifer's point of view her UE is local and Kristiina's is remote. Her UE now indicates its own preconditions line by line. Line 8:

```
a=curr:qos local none
```

states it has currently not reserved any local qos-related resources. Line 9:

```
a=curr:qos remote none
```

states it received information from the remote end (in the first offer) that no qos-related resources have been reserved at the moment. Line 10:

```
a=des:qos mandatory local sendrecv
```

states it mandatorily requires that its own resources get reserved in both the sending and receiving directions, before the audio session can start. Note that the initial value 'none', as set from the calling end, has changed to 'mandatory' because Jennifer's UE is also LTE attached and, therefore, is also mandated to reserve the resources locally before it can start sending media. Line 11:

```
a=des:qos mandatory remote sendrecv
```

Table 5.4 Policy rule example

SDP information	Policy rule information
Media type (audio)	QCI (1)
Speech codec (AMR)	Guaranteed bit rate/maximum bit rate (30 kbit/s)
IP addresses (Kristiina and Jennifer as communicated in c-lines of SDP)	Packet filter (5555::a:b:c:d and 5555::1:2:3:4)
Port numbers (Kristiina and Jennifer, as negotiated in Section 5.5.3	Packet filter (49152 and 49540)
a=inactive (flow status)	IP flow status (disabled, all media blocked)

states it received information from the remote end (in the first offer) that resources are mandatorily reserved in both the sending and receiving directions. Line 12:

```
a=conf:qos remote sendrecv
```

states the calling UE (remote) should send a confirmation (conf) at the moment the resources (qos) have been reserved in the sending and receiving directions (sendrecv). This is a new line that the called end adds to SDP. It is a necessary addition because the called UE is not intended to ring the called user or to start sending media until both ends have reserved the resources. This SDP answer is now sent in the 183 (Session Progress) response to the calling UE.

Once the P-CSCF receives the 183 response it passes the response towards Kristiina's UE but in addition it sends the diameter AA request (AAR) command to PCRF. This AAR command conveys session information (such as IP addresses, port numbers, media type (audio), negotiated codec(s) (AMR), bandwidth data, flow status (disabled as the session is not yet accepted) obtained from SDP offer/answer. When the PCRF receives this it performs QoS authorisation and finds the associated Gx session towards the S/P-GW using the served user's UE IP address. If the QoS authorisation is successful the PCRF creates policy rules based on information received in AAR (see Table 5.4) and sends the Rx AA answer (AAA; with a success value to indicate successful authorisation) towards P-CSCF.

The created policy rule is then delivered to P-GW using another diameter command, re-auth-request (RAR). This rule comprises information which is used by the P-GW to enforce QoS and to apply media flow and traffic treatment for voice. The policy and charging enforcement functionality (PCEF) inside the P-GW applies the received information which causes an establishment of a dedicated bearer for voice traffic, as shown in Figure 5.16.

1. The PCRF sends the RAR command towards the PCEF in the P-GW. This RAR contains media flow and QoS information, that is the flow information, QCI, maximum requested bandwidth UL/DL (MBR), GBR UL/DL, ARP for the media flows of the IMS voice session. Table 5.5 gives an example of the flow of QoS information, which comes from PCRF over the Gx reference point and is used by PCEF for dedicated bearer establishment.

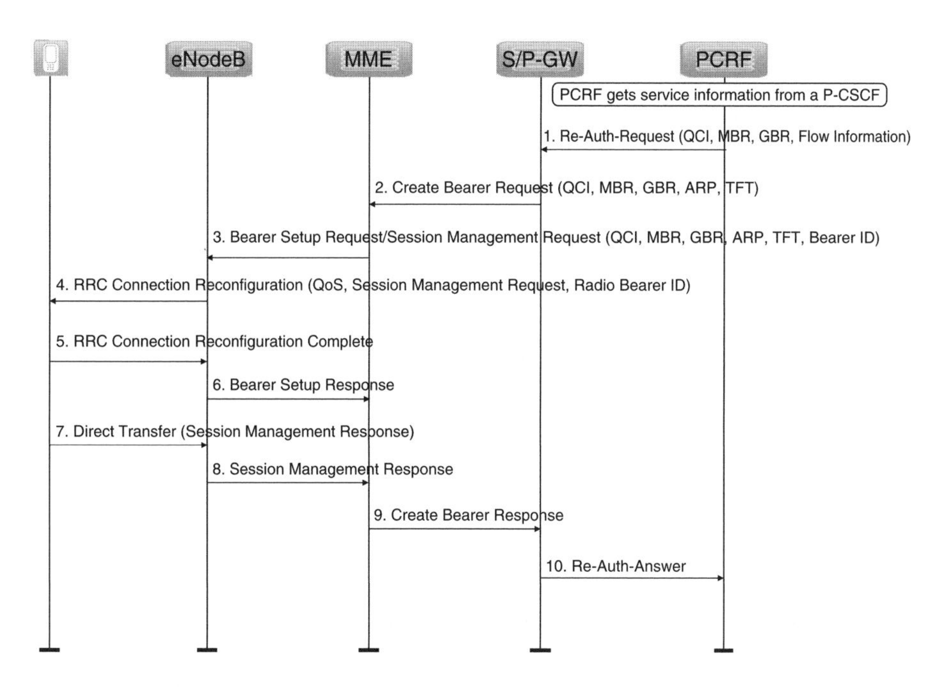

Figure 5.16 Dedicated EPS bearer establishment for voice media.

Table 5.5 QoS information over the Gx reference point

Protocol	IP
Source IP address	5555::1:2:3:4
Source port	49540
Destination port	49152
Destination IP address	5555::a:b:c:d
QCI	1
MBR (UL/DL)	30 kbit/s
GBR (UL/DL)	30 kbit/s

When the P-GW gets this information it recognises that there is no bearer available for the provided QCI and ARP pair. The P-GW initiates a new dedicated bearer creation. P-GW uses this flow and QoS information to establish the dedicated EPS Bearer. P-GW assigns bearer level values of QCI, ARP, GBR/MBR. In the example above the PCRF did not send ARP, so P-GW uses the default bearer ARP. The P-GW generates also a traffic flow templates (TFTs) based on the flow information. TFT is mandatory for the dedicated bearer and it is used by UE and P-GW to map traffic to right bearer.

2. The S/P-GW sends create bearer request messages to MME for establishing dedicated bearer which includes IMSI, EPS Bearer QoS, TFT and so on. For voice media the dedicated EPS bearer QoS has QCI = 1, MBR/GBR = 30 kbit/s and ARP = 9. TFT

contains uplink packet filters for UE, for example IP address = 5555::1:2:3:4 and port number = 49540.
3. The MME allocates EPS bearer identity for the dedicated bearer and sends it together with EPS bearer QoS and TFT information to the UE in the session management request inside the bearer setup request message to the eNodeB.
4. The eNodeB maps the EPS bearer QoS to the radio bearer QoS and then sends a RRC connection reconfiguration message to the UE. This contains also a session management request sent by MME.
5. The UE stores the new bearer QoS settings and EPS bearer identity given by the network in the session management request. Signalling contains also information to which default bearer this dedicated bearer is linked, called the linked EPS bearer identity (LBI). The UE uses uplink TFT to identify voice traffic flow coming from the application layer and matches uplink traffic to right radio bearer on the air interface. After configuration the UE returns the RRC connection reconfiguration complete messages to the eNodeB.
6. The eNodeB acknowledges the bearer activation to the MME with a bearer setup response message. The eNodeB indicates that requested EPS Bearer QoS can be allocated for voice media and there is 30 kbit/s GBR reserved.
7. The UE sends a session management response, including EPS bearer identity, to MME via eNodeB in a direct transfer message.
8. The eNodeB forwards the session management response message to the MME inside an uplink NAS transport message.
9. After radio setup the MME acknowledges the bearer activation to the S/P-GW by sending a create bearer response message.
10. Because the dedicated bearer was triggered by a PCRF, the P-GW sends a re-auth-answer (RAA) message to inform the PCRF of the successful rule installation.

After receiving the 183 (Session Progress) response Kristiina's UE acknowledges the receipt of 183 with a PRACK request. This request does not contain SDP information as the codec selection has been made and resources are not yet established; there is no change in SDP preconditions.

As soon as the P-GW initiated dedicated bearer (QCI1) is ready Kristiina's UE confirms this to the Jennifer's UE by sending a SIP UPDATE request. This UPDATE request will include a second SDP offer in its body, which shows that the resources at the calling end have been reserved:

```
m=audio 49152 RTP/AVP 97 98
a=sendrecv
a=rtpmap:97 AMR/8000/1
a=rtpmap:98 telephone-event/8000/1
a=curr:qos local sendrecv
a=curr:qos remote none
a=des:qos mandatory local sendrecv
a=des:qos mandatory remote sendrecv
```

The only differences here are that the first a=curr line has changed from 'none' to 'sendrecv' and the media status has changed from 'inactive' to 'sendrecv'. Consequently,

Kristiina's UE indicates that the status of the local QoS-related resources for the audio stream has changed. They have now been successfully reserved in both the sending and receiving directions. Let us assume that also Jennifer's UE has now a dedicated bearer established and it can immediately start to ring after sending the 200 (OK) response for the UPDATE request, because it is now sure that both ends have sufficient resources reserved to send and receive the audio stream. When it starts to ring, it sends a 180 (Ringing) response to the INVITE request in parallel.

The 200 (OK) for the update will include a second SDP answer with the following SDP information about the audio stream:

```
m=audio 49540 RTP/AVP 98 99
a=sendrecv
a=rtpmap:98 AMR/8000/1
a=rtpmap:99 telephone-event/8000/1
a=curr:qos local sendrecv
a=curr:qos remote sendrecv
a=des:qos mandatory local sendrecv
a=des:qos mandatory remote sendrecv
```

We can see from this that all the current states for resource reservation match the desired states; so, the preconditions negotiation has been successful and has finished.

5.5.5 Charging

This section explains how the communication services providers of our example users Kristiina and Jennifer perform their charging procedures. Here our assumptions are that both communication service providers are providing VoLTE service using the IMS APN and therefore all charging decision are done in the IMS layer and charging on IMS APN is disabled in EPC. Let us further assume that Kristiina is a postpaid (offline charging) subscriber and Jennifer is a prepaid (online charging) subscriber.

Figure 5.17 shows our example VoLTE session. It highlights key charging related SIP parameters and reveals when IMS functions contact online and offline charging entities which are further described in this section.

5.5.5.1 Offline Charging

Earlier in Section 5.5.2 we have shown that IMS signalling traverses through various IMS entities.[8] In fact, each offline charging-capable entity contains an integrated function called a charging trigger function (CTF). The CTF is aware of charging triggers (such as the beginning of IMS sessions, IMS session modification, IMS session termination, sending of message, subscribing to an event, publishing presence information) and is able to decide when it needs to contact the charging data function (CDF), the central point in the offline charging system. When a trigger condition is met the CTF collects charging

[8] Although, our example here shows only CSCFs the reader should keep in mind that all IMS entities are able to generate offline charging information as explained in Section 3.4.4.5.

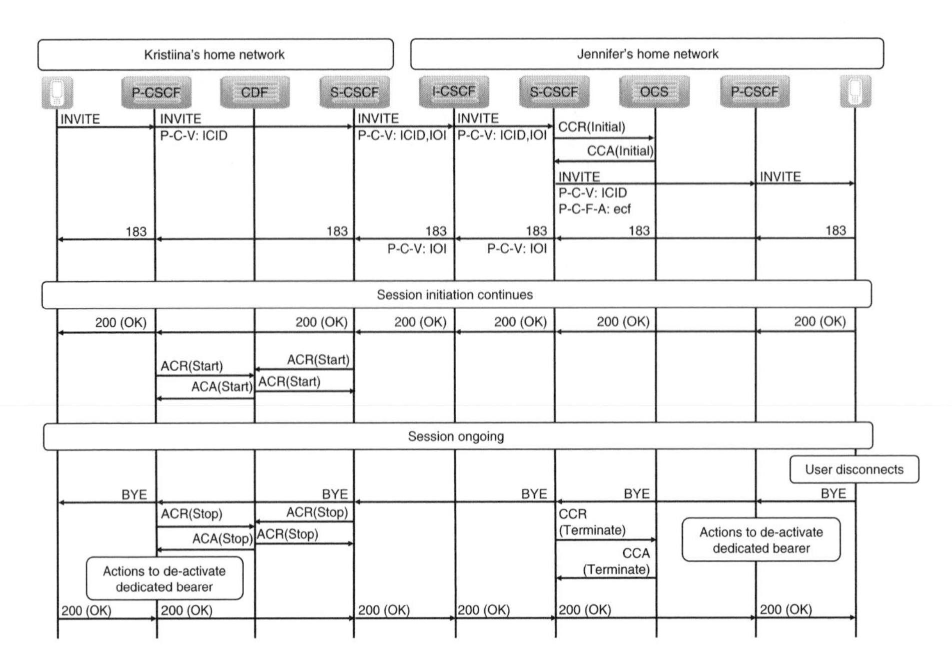

Figure 5.17 VoLTE charging.

information from the signalling message and sends the offline charging information to the CDF using diameter accounting requests (ACRs) via the Rf interface. The request contains much information about the event that launched the trigger (e.g. type of request INVITE/MESSAGE/SUBSCRIBE, calling party address, called party address, time stamps). The CDF uses a diameter accounting answer (ACA) to acknowledge the received request.

The offline charging system needs to support both session- and event-based charging. IMS functions know the request type that need to be indicated to the CDF. This is achieved using a suitable value in the accounting record type AVP ('event' for events and 'start', 'interim', 'stop' for sessions) in the ACR. IMS session-related ACRs are called start, interim and stop and are sent at the start, during and at the end of a session, as the name implies.[9] Nonsession-related ACRs are called event ACRs. Event ACRs cause the CDF to generate corresponding Charging Data Records (CDRs), while session ACRs cause the CDF to open, update and close the corresponding CDRs.

In an IMS session three different phases can be detected (session initiation, session modification, session release). At beginning of a session (200 OK, acknowledging an INVITE, is received), the CTF inside an IMS entity monitors the signalling traffic and detects a triggering point defined for reception of 200 OK, acknowledging an invite. When the triggering point is met the CDF collects information from the signalling messages (e.g. calling party address, called party address, time stamps, 'audio' SDP media component), assembles the charging information that matches the detected chargeable

[9] As an exception during the session setup phase I-CSCF may send ACR[Event] when it receives an INVITE request.

event and forwards the charging information towards the CDF via the Rf reference point using an ACR (start) request (see Figure 5.17). Usage of an ACR (start) prompts the CDF to open a CDR for this session. When the same session is modified (RE-INVITE or UPDATE is received), for example a video component is added – the CTF could again trigger this event and collect the necessary information again (e.g. calling party address, called party address, time stamps, 'audio + video' SDP media component). This changed charging information is sent again to the CDF, but this time an ACR (interim) request is used. Finally, when the session ends (BYE is received) the CTF constructs the ACR (stop) request to indicate session termination (see Figure 5.17). Based on these charging events, the CDF can create a single CDR including, for example A-party information, B-party information, total session time, audio session time and video session time.

In the case of a single end user to network transaction (e.g. sending a SIP MESSAGE containing SMS) a single ACR/ACA (event) is sufficient.

So far we have described how charging information is delivered from an IMS entity to the CDF. From Figure 3.22 we can see that there are further steps ahead before a billing system can send an actual bill to a user. The next step is to transfer CDRs from the CDF on towards the charging gateway function (CGF). The CGF is needed as there can be multiple CDFs involved in a single session/nonsession event because different IMS entities may send charging information to different CDFs (e.g. due to roaming or configuration reasons). The CGF validates, consolidates, pre-processes the incoming CDR (e.g. filters unnecessary fields and adds communication service provider-specific information), and may correlate different CDRs before passing them to the billing system. Table 5.6 summarises the key procedures supported by different offline charging functions.

5.5.5.2 Online Charging

The purpose of online charging is to perform credit control before usage of IMS services/resources. Two different models exist: direct debiting and unit reservation. In direct

Table 5.6 Summary of offline charging functions

Offline charging function	Key procedures
Charging triggering function (CTF)	Monitors SIP signalling Detects trigger condition Extracts information from SIP signalling and assemblies charing information Sends charing information to CDF
Charging data function (CDF)	Constructs CDRs Delivers CDRs to CGF
Charging gateway function (CGF)	Correlates, consolidates, filters unnecessary fields and adds communication service provider-specific information to the received account information CDR error handling and storage Delivers CDRs to billing system Pre-processes CDRs
Billing system	Creates the actual bill

debiting, an IMS network entity contacts the online charging system (OCS) and asks permission to grant the usage of services/resources. The OCS uses an internal rating function to find the appropriate tariff for an event based on the received information, if the cost was not given in the request. After resolving the tariff and the price, the OCS checks whether the user has enough credits in their account. If so, the OCS deducts a suitable amount of money from the user's account and grants the request from the IMS entity. In the unit reservation model, the OCS receives a credit control request from an IMS entity and uses an internal rating function to determine the price of the desired service according to the service-specific information provided by the IMS entity, if the cost was not given in the request. Then the OCS reserves a suitable amount of money from the user's account and returns the corresponding number of resources to the requesting IMS entity. Among the number of resources could be, for example, time or allowed data volume. When resources granted to the user have been consumed or the service has been successfully delivered or terminated, the IMS entity informs the OCS of the number of resources consumed. Finally, the OCS deducts the used amount from the user's account. It is also possible for the OCS to receive subsequent requests from the IMS entity during service execution if all granted resources are consumed. In this case the OCS needs to perform a new credit authorisation.

The direct debiting model is appropriate when the IMS entity knows that it could deliver the requested service to the user itself. For example, a game AS could send a credit control request and inform the OCS of the service (say, a game of rally) and the number of items (say, two) to be delivered. Then the OCS uses the rating function to resolve the tariff (0.3 euros) and to calculate the price based on the number of delivered units 0.6 euros). Finally, 0.6 euros are deducted from the user's account and the OCS informs the game AS that two units have been granted within the credit control answer. 3GPP's definition for this online charging model is immediate event charging (3GPP TS 32.240).

The unit reservation is suitable when the IMS entity is unable to determine beforehand whether the service could be delivered or when the required number of resources are not known prior to the use of a specific service (e.g. duration of multimedia session). The unit reservation model is usually applied to sessions (3GPP's term for this is session charging with unit reservation) but it is also possible to apply nonsession related requests (3GPP's term for this is event charging with unit reservation).

The Ro reference point is defined for online charging. It transfers credit control requests and answers between the OCS and three different IMS entities which are able to perform online charging [AS, multimedia resource function controller (MRFC) and S-CSCF via the IMS-gateway function (GWF)]. The credit control request and credit control answer commands from (RFC4006) are used for this purpose. In addition, 3GPP has defined 3GPP credit control AVPs to enhance the internet engineering task force's (IETF's) solution (3GPP TS 32.299) to meet the charging requirements of 3GPP.

To enable direct debiting the IMS entity sends a credit control request to the OCS and uses the value 'EVENT REQUEST' in a credit control request (CCR)-type AVP and the value 'DIRECT DEBITING' in the requested action AVP. For example, an IP short message gateway (IP-SM-GW) may receive a request from a user to send a message to somebody. The IP-SM-GW knows that the user is a pre-paid user and, therefore, it needs to seek permission from the OCS. It constructs a credit control request, sets the CCR-type AVP, requested action AVP and any other required AVP correctly, and then sends the request to the OCS. The OCS gets the request and, if it does not contain information about the price of service, the OCS uses a rating function before consulting the user's account. When the user has enough credits in the account the OCS grants the request with a credit control answer. Finally, the IP-SM-GW allows the service and sends the message towards the destination.

To enable session charging with unit reservation the IMS entity sends a credit control request to the OCS and then uses values 'INITIAL REQUEST', 'UPDATE REQUEST' and 'TERMINATION REQUEST' in the CCR-type AVP as follows:

- The 'INITIAL REQUEST' value is used when the IMS entity receives the first service delivery request.
- The 'UPDATE REQUEST' value is used when the IMS entity request reports the number of units used and indicates a request for additional units.
- The 'TERMINATION REQUEST' value is used when the IMS entity reports that the content/service delivery is complete or the final allocated units have been consumed.

To enable event charging with unit reservation the IMS entity sends a credit control request to the OCS and then uses the values 'initial request' and 'termination request' in the CCR-type AVP as follows:

- The 'INITIAL REQUEST' value is used when the IMS entity receives the first service delivery request.
- The 'TERMINATION REQUEST' value is used when the IMS entity reports that content/service delivery is complete.

5.5.5.3 VoLTE Charging

When Kristiina's P-CSCF receives the INVITE request it creates a new ICID (see Figure 5.17). With ICID it is possible to correlate session/transaction related charging data generated in different IMS elements. The ICID is included in the P-Charging-Vector (P-C-V) header.

```
INVITE sip:jennifer@csp.com SIP/2.0
P-Charging-Vector: icid-value="AyretyU0dm+6O2IrT5tAFrbHLso
     =023551024"
```

Kristiina's S-CSCF stores the ICID and adds an originating Inter Operator Identifier (IOI) in the P-C-V header. This IOI value reveals to Jennifer's communication service provider the originating communication service provider's identifier:[10]

```
INVITE sip:jennifer@csp.com SIP/2.0
P-Charging-Vector: icid-value="AyretyU0dm+6O2IrT5tAFrbHLso
     =023551024"; orig-ioi=example.home1.fi
```

When the INVITE request hits the I-CSCF it selects the S-CSCF and passes the request to the S-CSCF without modifying charging related information. Jennifer's communication

[10] The IOIs between P-CSCF and S-CSCF are exchanged during the IMS registration see Section 5.4.8.

service provider may define charging trigger also for the I-CSCF. If such a trigger is defined then the I-CSCF sends an ACR (event) to CDF which creates an I-CSCF CDR right away.

Jennifer's S-CSCF has downloaded subscription data during her registration and this data contains information that Jennifer is an online subscriber and that is why S-CSCF needs to interact with OCS. To enable session charging with unit reservation the S-CSCF sends a credit control request to the OCS and uses value 'INITIAL REQUEST' to indicate that the S-CSCF has received the first service delivery request. The OCS receives the request and uses the information provided [e.g. SDP media information (voice) and IMS communication service identifier, ICSI (MMtel)] to decide whether or not to grant this request. The credit control answer contains the number of service units granted (e.g. 10 min) and based on this the S-CSCF is able to allow the SIP session to continue. After credit control check the S-CSCF will store and remove the received orig-ioi parameter from the P-C-V header and it will send the INVITE request to the P-CSCF:[11]

```
INVITE sip:[5555::1:2:3:4]:1600 SIP/2.0
P-Charging-Vector: icid-value="AyretyU0dm+6O2IrT5tAFrbHLso
     =023551024"
P-Charging-Function-Addresses: ecf=onlinechargingentity.example.com
```

The P-CSCF of Jennifer will store the charging information and it will remove all information before sending the invite request to the UE.

When receiving the first response, that is 183 (Session Progress), Jennifer's P-CSCF will add the P-C-V header, including the same ICID value as received in the invite request:

```
SIP/2.0 183 Session Progress
P-Charging-Vector: icid-value="AyretyU0dm+6O2IrT5tAFrbHLso
     =023551025"
```

Jennifer's S-CSCF will add the terminating IOI information before sending it further. This IOI value reveals to Kristiina's communication service provider the terminating communication service provider's identifier:

```
SIP/2.0 183 Session Progress
P-Charging-Vector: icid-value="AyretyU0dm+6O2IrT5tAFrbHLso
     =023551025"; term-ioi=csp.com
```

The terminating IOI information will be stored and removed by Kristiina's S-CSCF and her P-CSCF will remove the P-C-V header from the response.

Eventually Kristiina's S-CSCF receives the 200 (OK) response to the INVITE request. This will trigger S-CSCF to send a diameter ACR (start) command to CDF. The CDF will open S-CSCF CDR based on received information and it acknowledges the request with an ACA command. Similarly the P-CSCF sends a diameter ACR (start) command to CDF.

[11] According to (3GPP TS 24.229) the INVITE request to the P-CSCF also contains P-C-F-A header containing the address of OCS. The P-CSCF does not need this information as it is not capable for online charging.

The CDF will open P-CSCF CDR based on received information and it acknowledges the request with ACA command. Both S-CSCF and P-CSCF will use single CDF as the S-CSCF has delivered the address of CDF during Kristiina's registration (see Section 5.4.8)

Once the Jennifer ends the session the BYE request triggers Jennifer's S-CSCF to send a CCR (termination) command to the OCS informing that the service has been successfully terminated and it also delivers information regarding consumed resources. Based on the unconsumed resources the OCS updates Jennifer's account.

The S-CSCF and P-CSCF of Kristiina will send an ACR (stop) command to the CDF which in turn closes the CDRs and eventually the CDF delivers information to the billing system via the CGF. The P-CSCF further informs the PCRF that session has been terminated and the PCRF in turn will remove policy rules from the P-GW and finally P-GW de-activates the dedicated bearer used for IMS user plane traffic.

5.5.6 Session Release

5.5.6.1 User-Initiated Session Release

When user wants to stop their conversation and presses the red button then the UE generates a BYE request, which is sent to the other end in the same way as any other subsequent request (as shown in Figure 5.18). In parallel to this, the PCRF will also remove dedicated bearer authorisation and consequently the P-GW will initiate bearer deactivation.

Let us assume that Jennifer drops the session first and her UE sends the BYE request:

```
BYE sip:[5555::a:b:c:d]:1400 SIP/2.0
Route:<sip:pcscf2.csp.com:1500;lr>
Route:<sip:scscf2.csp.com;lr>
Route:<sip:scscf1.example.com;lr>
Route:<sip:pcscf1.example.com;lr>
To:<sip:kristiina@example.com>;tag=171828
From:<sip:jennifer@csp.com>;tag=333333
```

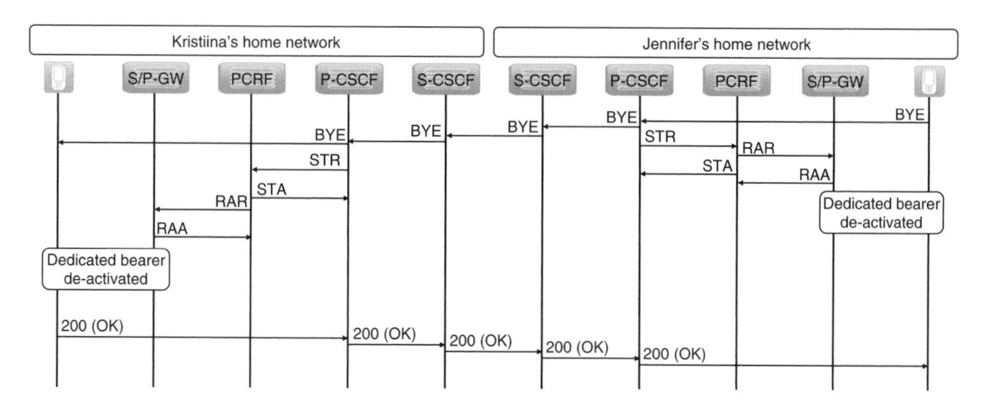

Figure 5.18 UE initiated session release.

Once the Kristiina's UE receives the BYE request it will also respond to the request with a 200 (OK) response, which will be sent back towards Jennifer. The four CSCFs on the route will clear all dialogue states and information related to the session. P-CSCFs on the route will send session termination request (STR) diameter commands to inform the PCRFs that session that was previously authorised is now ended and the PCRF shall enforce that associated bearer resources will be released in the P-GWs to prevent bearer misuse after SIP session termination. The PCRF acknowledges the received STR command with a session termination answer command and it will send a RAR command to P-GW to remove previously installed policy rule (see Section 5.5.4). The P-GW removes the identified policy rule and detects that no rules are remaining the bearer and P-GW starts deactivation of the dedicated bearer(s). P-GW sends a RAA command to acknowledge the RAR command and informs the PCRF about the removed rules.

5.5.6.2 Network-Initiated Session Release

There might be situations in which it is necessary for one of the CSCFs to release the session, rather than the user. For example, Jennifer's P-CSCF would need to release an ongoing session when it realises that Jennifer's UE has lost the bearer transporting the IMS user plane traffic, for example due to lost radio coverage (Figure 5.19).

In that case the P-GW reports the incident to the PCRF using a CCR command. The PCRF acknowledges this with a CC answer (CCA) command and it also sends an abort session request (ASR) command to the P-CSCF to inform that the user plane bearer related to ongoing SIP session has been released. The P-CSCF responses with an abort session answer (ASA) command and sends a bye request on behalf of Jennifer:

```
BYE sip:[5555::a:b:c:d]:1400 SIP/2.0
Route: <sip:scscf2.csp.com;lr>
Route: <sip:scscf1.example.com;lr>
Route: <sip:pcscf1.example.com;lr>
To: <sip:kristiina@example.com>;tag=171828
From: <sip:jennifer@csp.com>;tag=333333
```

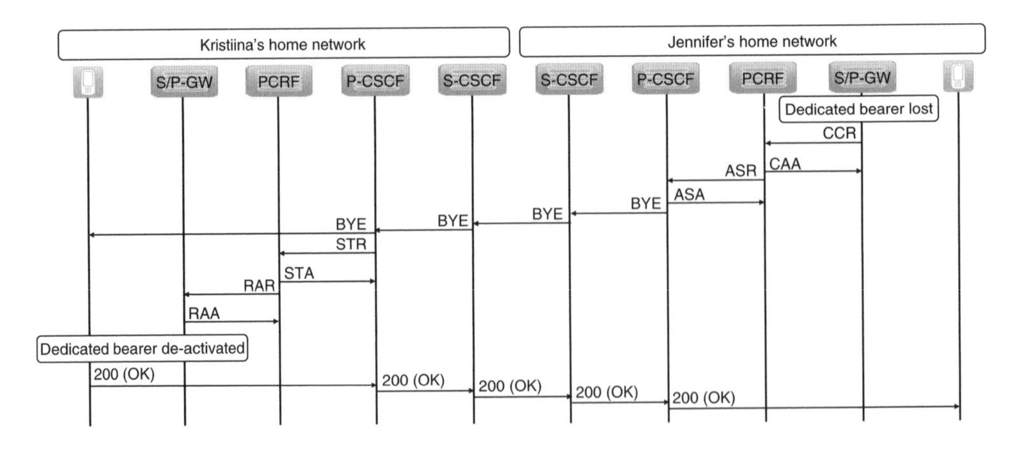

Figure 5.19 Example of network initiated session release.

There may be occasions when Kristiina's S-CSCF needs to be shut down or Kristiina may be using a pre-paid card and runs out of money. In such cases Kristiina's S-CSCF would release the session by issuing one BYE request towards Kristiina's UE:

```
BYE sip:[5555::a:b:c:d]:1400 SIP/2.0
Route: <sip:pcscf1.example.com;lr>
To: <sip:kristiina@example.com>;tag=171828
From: <sip:jennifer@csp.com>;tag=333333
```

and another bye request towards Jennifer's UE:

```
BYE sip:[5555:1:2:3:4]:1600 SIP/2.0
Route: <sip:scscf2.csp.com;lr>
Route: <sip:pcscf2.csp.com;lr>
From: <sip:kristiina@example.com>;tag=171828
To: <sip:jennifer@csp.com>;tag=333333
```

To generate the BYE request with the correct set of Route headers, the P/S-CSCFs need to keep track of all routing information that is collected during the establishment of any dialogue.

5.5.7 Related Standards

Specifications relevant to Section 5.6 are:

- 3GPP TS 23.203 – Policy and charging control architecture.
- 3GPP TS 23.218 – IP multimedia (IM) session handling; IM call model.
- 3GPP TS 23.228 – IP multimedia subsystem (IMS); stage 2.
- 3GPP TS 24.173 – IMS multimedia telephony service and supplementary services.
- 3GPP TS 24.229 – IP multimedia call control protocol based on session initiation protocol (SIP) and session description protocol (SDP); stage 3.
- 3GPP TS 24.930 – Signalling flows for the session setup in the IP multimedia core network subsystem (IMS) based on session initiation protocol (SIP) and session description protocol (SDP); stage 3.
- 3GPP TS 26.114 – IMS multimedia telephony; media handling and interaction.
- 3GPP TS 29.212 – Policy and charging control over Gx reference point.
- 3GPP TS 29.213 – Policy and charging control signalling flows and quality of service (QoS) parameter mapping.
- 3GPP TS 29.214 – Policy and charging control over Rx reference point.
- 3GPP TS 32.260 – Telecommunication management; charging management; IP multimedia subsystem (IMS) charging.
- 3GPP TS 32.299 – Telecommunication management; charging management; diameter charging applications.
- RFC2833 – RTP payload for DTMF digits, telephony tones and telephony signals.
- RFC3261 – SIP: session initiation protocol.
- RFC3262 – Reliability of provisional responses in the session initiation protocol (SIP).

- RFC3264 – An offer/answer model with SDP.
- RFC3311 – The session initiation protocol (SIP) update method.
- RFC3312 – Integration of resource management and session initiation protocol.
- RFC3323 – A privacy mechanism for the session initiation protocol (SIP).
- RFC3325 – Private extensions to the session initiation protocol (SIP) for asserted identity within trusted networks.
- RFC3455 – Private header (P-header) extensions to the session initiation protocol (SIP) for the 3rd generation partnership project (3GPP).
- RFC3550 – RTP: a transport protocol for real-time applications.
- RFC3551 – RTP profile for audio and video conferences with minimal control.
- RFC3556 – SDP bandwidth modifiers for RTCP bandwidth.
- RFC3840 – Indicating user agent capabilities in the session initiation protocol (SIP).
- RFC3841 – Caller preferences for the session initiation protocol (SIP).
- RFC4566 – SDP: session description protocol.
- RFC4596 – Guidelines for usage of the session initiation protocol (SIP) caller preferences extension.
- RFC5279 – A uniform resource name (URN) namespace for the 3rd generation partnership project (3GPP).
- http://www.3gpp.org/tb/Other/URN/URN.htm – URN values maintained by 3GPP.
- RFC6050 – A session initiation protocol (SIP) extension for the identification of services.

5.6 Voice Continuity

5.6.1 PS-PS Intersystem Handover

In future there will come also IMS VoIP over HSPA, which means that the IMS VoIP capable terminal is able to move between LTE and HSPA accesses over the packet switched network during a VoIP call. For seamless handovers the network and UE must support packet switched handover (PS HO) where the network makes handover preparation before the actual inter-system change.

3GPP has specified two inter system mobility variants, Gn and S3 based on the used interfaces between MME and SGSN. Both variants offer an optimised inter-system PS handover from LTE to 3G (and vice versa). When PS HO is used the handover delay and data transfer interruption time are minimised as the handover preparation is done before the execution.

The PS handover procedure can also be used with CSFB or with SRVCC. When PS HO is done with CSFB it reduces a voice call setup time because the time needed for the access change is minimised. You can find more about CSFB in Sections 4.3.2 and 5.8.

For SRVCC the MME makes bearer splitting. The MME signals voice media bearer information to the MSC and all other bearers to the SGSN. The PS HO can be used for the MME-SGSN signalling to reduce a PS bearer handover delay. A more detailed SRVCC description is in Section 5.6.2.

The PS-PS handover is usually explained in two phases: handover preparation and handover execution. Figure 5.20 gives a high level description how handover preparation

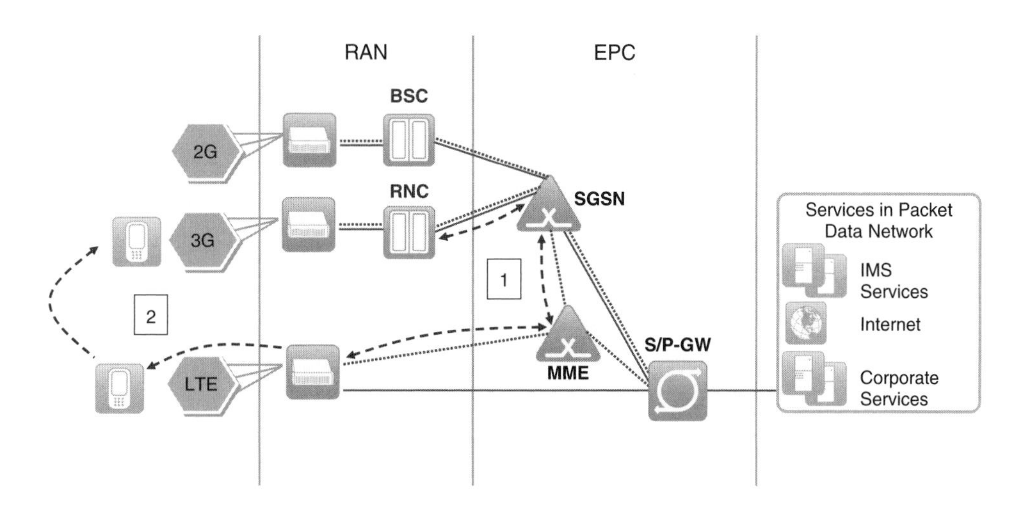

Figure 5.20 E-UTRAN to UTRAN PS-PS handover.

is done between the LTE eNodeB and 3G radio network controller (RNC), and core network signalling goes over an S3 interface between MME and SGSN.

1. The PS-PS VoIP handover from LTE to UTRAN is done so that ongoing IMS signalling bearer and VoIP media bearer are moved from source access to the target access by establishing corresponding bearers to the target access. In the handover preparation phase the target SGSN makes one-to-one mappinng from the EPS bearers to PDP contexts. The SGSN makes also mapping between the EPS Bearer QoS parameter values and the UTRAN QoS parameter values of a PDP context as shown in Figure 8.3 in Chapter 8. The target RNC allocates the resources for the voice call and returns the applicable parameters to SGSN, which responses to MME.
2. In the handover execution the MME sends a handover command to eNodeB including the bearer information. The eNodeB commands the UE to make handover to the target access network and handover execution is completed in the target side. After successful handover the MME starts releasing bearers on the source side.

5.6.2 Single Radio Voice Call Continuity

Another flavour to provide session continuity in VoLTE is to perform transfer of ongoing voice call from LTE (EPS/IMS domain) to circuit switched domain in case native VoIP connection no longer can be maintained in LTE. This procedure is standardised under name SRVCC but also known as packet switched–circuit switched (PS-CS) access domain transfer. The overall behaviour of VCC is based on principles defined in 3GPP Release 7 (voice call continuity) and further enhanced in 3GPP Release 8 in order to support network-initiated behaviour triggered by the access network. Additionally SRVCC has been defined in 3GPP Release 8 for both VoIP sessions started in UTRAN and in LTE but practical implementations can also be done only for LTE if so required.

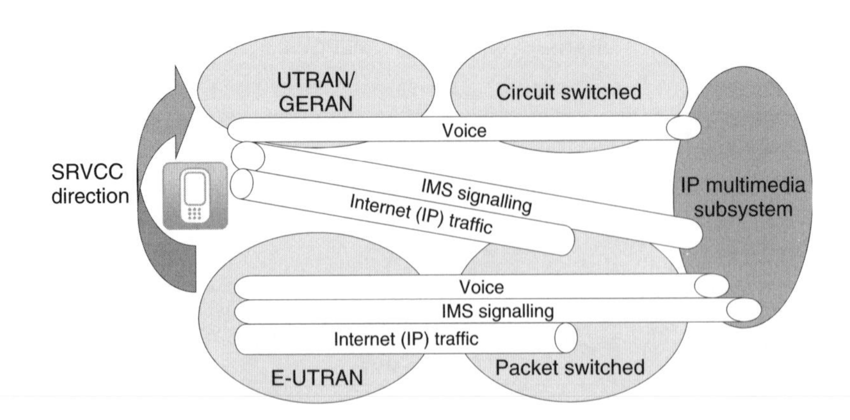

Figure 5.21 High level principle of single radio voice call continuity.

Figure 5.21 represents the high level idea of this procedure. The voice part of communication between UE and network is transferred from LTE to CS whereas other communication is maintained via the PS core.

The domain transfer consists at high level both moving the voice bearer from EPS to circuit switched domain but also moving related packet switched connection by using, for example PS handover procedure in parallel to the target radio access technology such as 3G.

In order to support SRVCC at system level specific new functionalities are required into network architecture. First of all new interworking functionality is defined as a SRVCC enhanced MSC server that is able to perform required procedures towards circuit switched domain to ensure that target radio access is prepared and ready for access domain transfer. Additionally EPS need to be enhanced, namely MME and evolved NodeB (eNb) to support suitable procedures to trigger domain transfer and eventually also move related packet switched EPS bearers besides the voice connection to the target access domain.

In order to support access domain transfer, the SRVCC enhanced MSC server has specific GPRS tunnelling protocol (GTP) based Sv-interface towards MME. This Sv interface as defined in (3GPP TS 29.280) is used by MME to request the MSC server to reserve required radio access resources from target CS radio access (GERAN/UTRAN) for SRVCC. Role of the SRVCC enhanced MSC server is twofold. First of all it is responsible to prepare resources towards IuCS or A interface either controlled by same network element or alternatively in case target radio access is controlled by another MSC Server (MSC-B), perform the normal inter-MSC relocation as defined in (3GPP TS 23.009). Secondly, after target CS radio access resources have been committed then the SRVCC enhanced MSC server will establish call on behalf of UE to a specific address given by MME via the Sv interface. This address is related to current service centralisation and continuity (SCC) AS of that particular subscription and involved in the original call establishment.

SRVCC procedure has been gradually improved since 3GPP Release 8 to support more functionality such as capability to support SRVCC for emergency calls, multiple simultaneous calls which are active and held, for calls in ringing phase, for video calls as well as capability to perform SRVCC back from CS network to LTE. In order to support functionalities beyond 3GPP Release 8 additional requirements set by 3GPP ICS

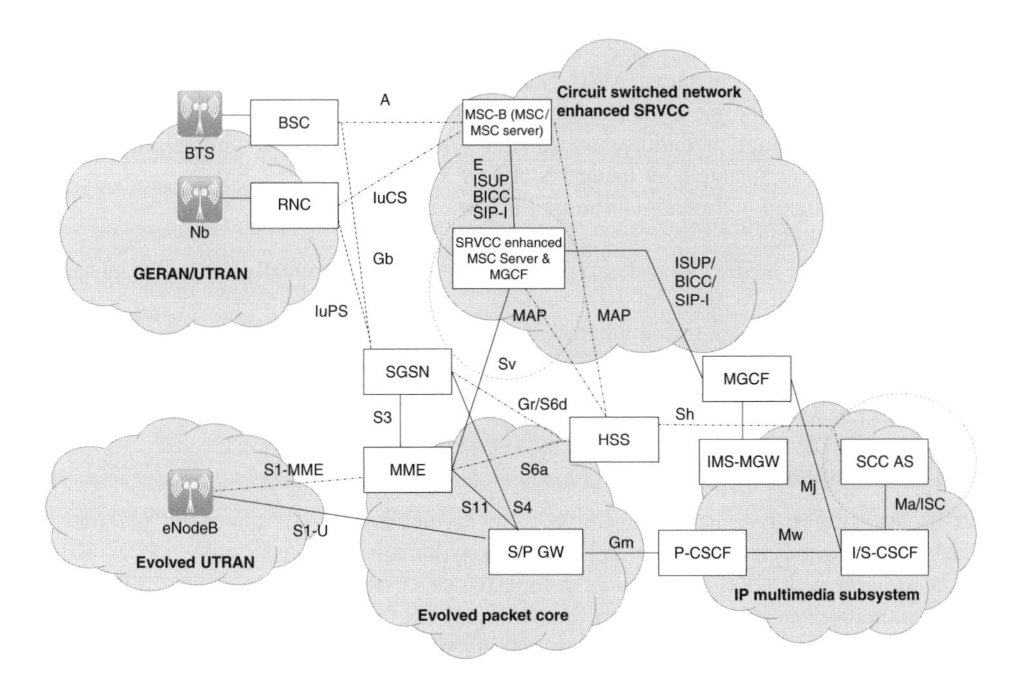

Figure 5.22 Network architecture for SRVCC.

architecture need to be taken into use. This can occur in a phased manner, in the case in which IMS-based voice over LTE has been deployed commercially by using 3GPP Release 8 standardisation baseline, which is also covered more thoroughly in this book.

Figure 5.22 represents a 3GPP Release 8-based network architecture required to perform SRVCC procedure.

In summary, certain relevant functionalities can be highlighted in this architecture for SRVCC access domain transfer execution:

- SRVCC enhanced MSC server:
  - Performs the task of preparing the target radio access side resources for immediate domain transfer.
  - Provides the Sv interface towards EPC, which will be used to trigger domain transfer.
  - Acts as anchor MSC (MSC-A) for the CS handover connection towards target MSC (MSC-B) in the case when inter-MSC handover occurs.
  - Updates the remote side of session after domain transfer.
- SCC AS:
  - Anchors the session during setup phase so that domain transfer can be later performed, if required.
  - Performs access transfer from one domain to another.
  - Performs terminating access domain selection which is elementary procedure to route terminating calls to a VoLTE subscriber.
  - Is able to remove media which is not allowed to continue after domain transfer.
- MME:
  - Indicates to eNb whether SRVCC is possible for this session.

  - Is responsible to find right EPS bearer that will be subject for domain transfer.
  - Invokes SRVCC enhanced MSC Server via Sv interface and provides sufficient information to complete the domain transfer procedure.
  - Performs relevant procedures to transfer non-VoIP bearers to target radio access technology for instance by using PS HO, if applicable.
  - Handle suspend and deactivation of EPS bearer when needed.
- E-UTRAN eNodeb:
  - May generate a SRVCC specific neighbour cell list dynamically for target measurement purpose
  - Is responsible to determine whether domain transfer needs to be performed.
  - Invokes MME to start domain transfer procedure via the S1 interface.
  - Invoke MME to start PS-HO in addition to PS to CS HO or just PS to CS HO

5.6.2.1 Execution of SRVCC

Let us use an already familiar example from earlier chapters of this book in order to clarify required the procedures to perform access domain transfer.

Let us assume that Kristiina and Jennifer have successfully established VoLTE session between themselves and are currently discussing a new fancy car that Kristiina has bought during the weekend. Kristiina is currently talking while driving (naturally using her Bluetooth headset) to the countryside from the city centre and, since her favoured network service provider LTE coverage has not yet reached the countryside where she is driving, the network decides (based on measurement data received from her smartphone) that a need for access domain transfer has appeared.

This starting point is shown in the Figure 5.23.

The dashed line illustrates the path taken by the SIP during call establishment phase. Services indicate both SCC AS and supplementary service (TAS) services but also other services may have been applied for the session. The solid line illustrates path taken by the user plane (i.e. speech) connection to other side of VoLTE session. In this example it carries voice between Jennifer and Kristiina. Finally the LTE UE typically also has many other applications that are connected to Internet services (dotted line), which use need to be able to continue after access domain transfer.

In order to start the access domain transfer procedure, we follow the flow shown in Figure 5.24 which describes the number of different signalling procedures that need to take place in order to securely transfer Kristiina's call from LTE radio access to 3G radio access available in the countryside.

This PS to CS access domain transfer procedure is triggered by the eNb, which performs this decision based on received measurements from UE. As a prerequisite, UE has an ongoing IMS session and a related dedicated EPS bearer to IMS APN with GBR nature; then eNb, based on pre-configured triggering condition, is able to recognise which EPS bearers are subject for the access domain transfer and which EPS bearers will need normal packet switched mobility towards target radio access network. The SR-VCC is typically triggered by UE event triggered measurements when the signal level [reference symbol received power (RSRP)] or the signal quality [reference signal received quality (RSRQ)] drops below a predefined threshold. When eNodeB receives the measurement report, it will initiate the SR-VCC procedure. In parallel to the SR-VCC procedure MME will also perform PS HO to GERAN/UTRAN since LTE attached UE has always at

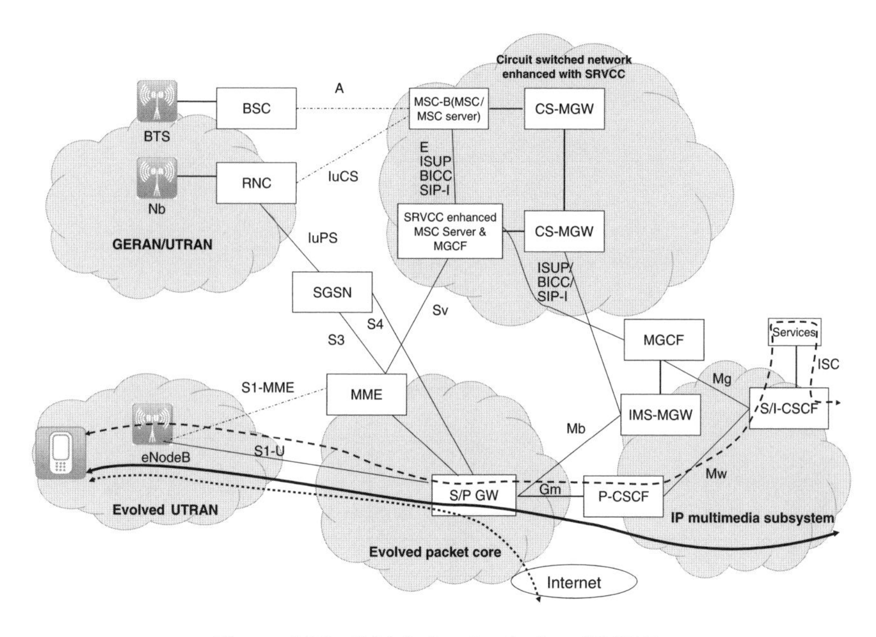

Figure 5.23 Initial situation before SRVCC.

least a default EPS bearer connection active. For more information about PS HO, see also Section 5.6.1.

In order to recognise which EPS bearer(s) needs to be transferred as SRVCC and which as PS HO, MME need to be aware that which EPS bearer has a voice component and which ones are dedicated for nonvoice use of those EPS bearers using the IMS APN. This is made possible with the use of 3GPP PCC architecture, as defined by (3GPP TS 23.203), in such a way that P-CSCF will request PCRF to provide QoS in form of an EPS dedicated bearer with a GBR and QoS class identifier value 1 (Voice), see Section 5.5.4. The SRVCC is applicable only when the user has a dedicated bearer with QCI 1 and based on this information the MME will pick the right EPS bearer from other active EPS bearers.

In the case where the target radio access technology is GERAN then the continuation of data connection after SRVCC requires the use of a dual transfer mode (DTM) feature in GERAN since otherwise ongoing GBR data bearers will be suspended for the duration of time and non-GBR data bearers will be released when UE is attached to GERAN and has an ongoing call. In the case where the target radio access technology is UTRAN then the multi-radio access bearer (RAB) functionality of UTRAN ensures a continuation of data connection similar to today.

MME start to prepare resources towards target networks, which also include a circuit switched domain in addition to a traditional packet switched domain since domain transfer is required. Based on local configuration MME will select a SRVCC enhanced MSC server having an Sv interface but also selects a target legacy packet switched network, namely SGSN, which MME then needs to contact via an S3 interface.

After this MME will send the following SRVCC PS to CS request message (Table 5.7) to the SRVCC enhanced MSC server via the Sv interface which again will trigger reservation of the target CS network resources needed to perform domain transfer.

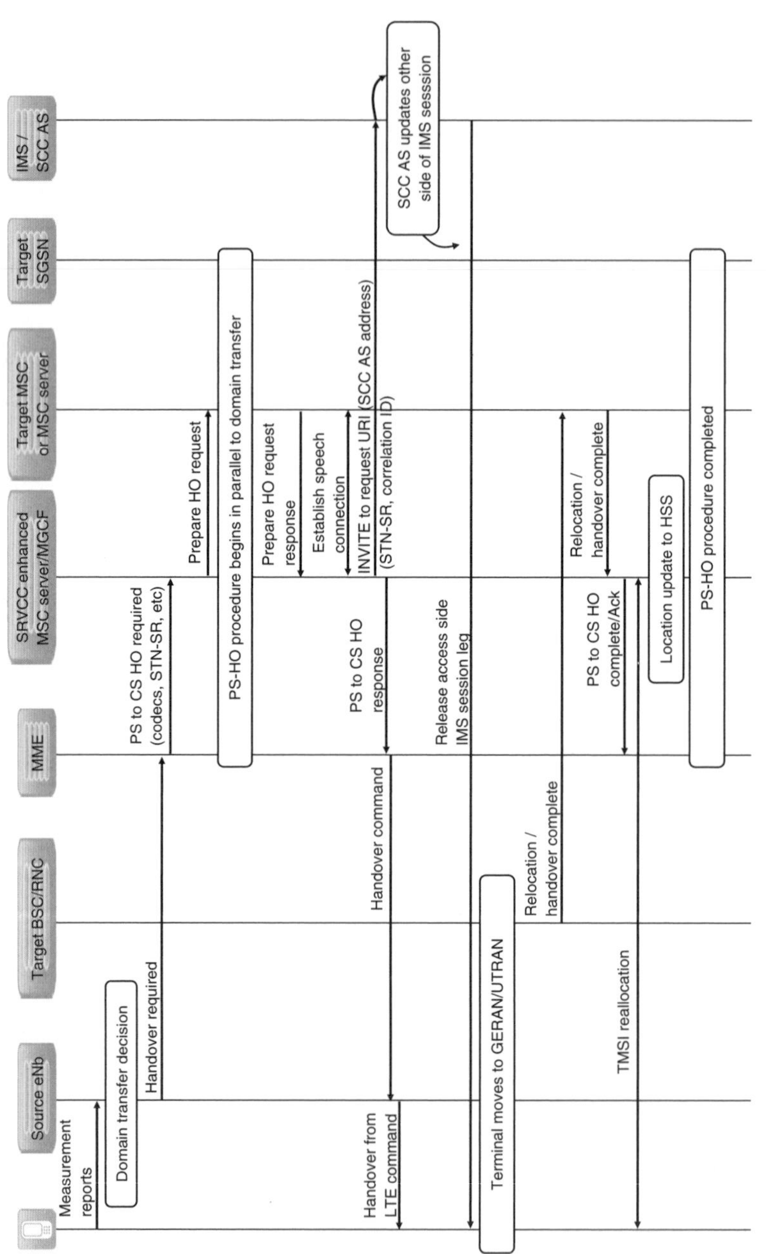

Figure 5.24 SRVCC signalling flow.

Table 5.7 Summary of SRVCC PS to CS request message

Name of parameter	Purpose	Example value/content
IMSI	This parameter identifies the subscriber performing the domain transfer.	Identifier of subscription (15 digits long), including mobile country code, mobile network code and actual mobile subscriber identity number (MSIN).
	SRVCC enhanced MSC Server uses this after successful domain transfer to perform location update to HSS.	
Mobile equipment identity (IMEI)	This parameter is only present if domain transfer will be performed without prior authentication or if UE is USIMless. In other words it is true IMSI is not available.	Identifier (16 digits long), containing eight-digit type allocation code (TAC) identifying the terminal model, six-digit serial number (SN) and two-digit software version (SV).
	This parameter is relevant in the case of SRVCC for an emergency call.	
Sv flags	This parameter is used to indicate whether the nature of a transferred session is an emergency call or in the case when the subscriber uses ICS services.	EmInd = 0/1
	ICS indication is received from HSS via S6a interface by MME and is provided to SRVCC enhanced MSC server in this parameter.	ICS = 0/1
MME/SGSN Sv address for control plane	This parameter contains the IP address used by MME to handle this domain transfer.	Usable IP address, such as 192.168.1.10
MME/SGSN Sv TEID for control plane	This parameter contains the tunnel endpoint identifier (TEID) value used by MME to handle this domain transfer. The same value is needed in GTP messages sent and received by network elements related to this domain transfer procedure.	Unique identifier value, such as 123
C-MSISDN	This parameter contains a unique address that identifies the end user and can be used for correlation purposes in SCC AS.	+3585012345678
	In the case of an emergency call without prior authentication or using USIMless UE then this parameter does not exist.	

(*continued overleaf*)

Table 5.7 (*continued*)

Name of parameter	Purpose	Example value/content
STN-SR	This parameter contains a session transfer number for SRVCC. Parameter is received by MME from HSS via S6a interface and uniquely identifies the SCC AS that has been used to anchor all sessions of a given subscriber.	+35850987654321
MM context for E-UTRAN SRVCC	This parameter contains the required information from the source side of a domain transfer to the target side of the domain transfer in order to perform the domain transfer. This parameter is only exchanged when a domain transfer is performed from E-UTRAN to UTRAN/GERAN, that is when SRVCC is performed for VoIP session over E-UTRAN. Information included for instance: mobile station class marks, supported codec list, security parameters.	Encoding of mobile station classmarks (2/3) as well as supported codec list is defined in (3GPP TS 24.008). Encoding of security parameters (CK_{SRVCC}, IK_{SRVCC}, and eKSI) as defined in (3GPP TS 33.401) and (3GPP TS 24.301).
MM context for UTRAN SRVCC	This parameter carries the required information from the source side of a domain transfer to the target side in order to perform a domain transfer. This parameter is only exchanged when a domain transfer is performed from UTRAN to UTRAN/GERAN, that is when SRVCC is performed for a VoIP session over UTRAN. Information include for instance: mobile station class marks, supported codec list, security parameters.	Encoding of mobile station classmarks (2/3) as well as supported codec list is as defined in (3GPP TS 24.008). Encoding of security parameters (CK'_{CS}, IK'_{CS}, KSI'_{CS}, Kc', $CKSN'_{CS}$) as defined in (3GPP TS 33.102) and (3GPP TS 24.008).
Source to target transparent container	This parameter is transferred from source eNb/Nb transparently to target Nb/base transceiver station (BTS) via core network.	When target radio access network is GERAN, this container contains old base station system (BSS) for the new BSS information element as defined in (3GPP TS 48.008).

Table 5.7 (*continued*)

Name of parameter	Purpose	Example value/content
	It contains important information for the target radio access network to establish sufficient radio resources for the transferred call.	When the target radio access network is UTRAN, this container contains a source RNC to target RNC transparent container as defined in (3GPP TS 25.413).
Target RNC ID	This parameter identifies the target RNC.	As described in 3GPP TS 29.002.
Target cell ID	This parameter identifies the target cell.	As described in 3GPP TS 24.008.
Source (SAI)	This parameter identifies the source Service Area Identifier (3G Cell) and is only valid if the SRVCC is from HSPA.	As described in 3GPP TS 24.008.

The Table 5.7 explains the relevant details of the SRVCC PS to CS request message required to trigger domain transfer properly. This message has been specified in (3GPP TS 29.280) at stage 3 level.

Next the SRVCC enhanced MSC server will verify the ownership of the target RNC or base station controller (BSC) based on the local configuration data (radio access network configuration) and as result of this task will have the identity of the target MSC/MSC server network element to which the handover will be performed. Two alternatives exist. In the case where target CS radio access is controlled by the same MSC server network element which acts as an SRVCC enhanced MSC Server then it locally prepares resources towards RNC or BSC. However in the case where inter-MSC relocation is needed within the scope of the SRVCC procedure, the SRVCC enhanced MSC server shall contact the MSC or MSC server that controls the target BSC or RNC via mobile application part (MAP) based e-interface (i.e. MSC-B) as in the case of inter-MSC handover today.

When the required resources have been reserved from the target circuit switched radio access network (UTRAN or GERAN) the SRVCC enhanced MSC server uses the session transfer number for SRVCC (STN-SR) and routes the call towards the anchor functionality that resides within the IMS architecture, that is SCC AS that has anchored the IMS session initially during the session establishment phase. This address, which uniquely identifies the network element in the home IMS network which previously anchored the given session for a particular subscriber (in this case Kristiina), is obtained from HSS by MME during the LTE attachment phase and given via the Sv interface to an SRVCC enhanced MSC server as described earlier. STN-SR address allocation needs to be aligned at system level to the IMS configuration (initial filtering criteria) since SCC AS selection during original call establishment, as described earlier, has been performed by using initial filter criteria (iFC) stored in subscription data in the HSS.

Additionally the SRVCC enhanced MSC server contains uses the correlation mobile subscriber international ISDN number (C-MSISDN) identifying the subscriber (Kristiina) that is the subject for the domain transfer and is received from MME via the PS-CS handover message. The SCC AS will use this C-MSISDN address to associate the incoming session [via the media gateway control function (MGCF)] from the SRVCC enhanced MSC server to the already ongoing anchored IMS session. Additionally SCC AS will immediately update another side of the IMS session with a new local descriptor having a new IP address and port number that should be used for RTP and RTCP sessions from this point onwards by another side of the IMS session (Jennifer's UE). Additionally SCC AS clears the SIP signalling connection towards UE that is performing the domain transfer (Kristiina's).

The IP address and port number given by SCC AS to the another side of connection belong to the MGW and which are sent by SRVCC enhanced MSC server within the SDP offer of a particular session transfer call routed by using the STN-SR address to the SCC AS.

The following example represents the content of the invite sent by the SRVCC enhanced MSC server, also having in this example a MGCF role as integrated functionality (3GPP TS 29.163) and thus simplified for the sake of clarity. Only the relevant parameters are shown. In order to have a better understanding about the use of SIP in the context of VoLTE, the reader is advised to read Section 5.5.

```
INVITE tel:+358040987654321 SIP/2.0
P-Charging-Vector: icid-value="GseadArs+6OLdwdoiDKLoo=0412300";
orig-ioi= example.home1.fi
P-Asserted-Identity: tel:+358501234567

. . .
Content-Type: application/sdp
Content-Length: (. . .)

m=audio 49100 RTP/AVP 97
c=IN IP6 5555::aaa:bbb:ccc:eee
a=rtpmap:97 AMR/8000/1
a=rtpmap:99 telephone-event/8000/1
```

The actual content of the message is similar to normal SIP session establishment with the following exceptions:

- The INVITE request's request URI contains a routable E.164 address of SCC AS that has been retrieved via MME from HSS by the SRVCC enhanced MSC Server.
- The P-C-V is populated by the SRVCC enhanced MSC Server and is different from the one originally created for the session. This header is used for correlating the charging information in the post-processing system and therefore both the old and this new value need to be given together by SCC AS to the post-processing system in order to correlate this session transfer call into the correct session.
- The P-asserted-identifier includes a correlation MSISDN address received via MME from HSS by the SRVCC enhanced MSC Server and identifies the MSISDN of the subscriber who currently has performed domain transfer (Kristiina in this example).
- The SDP contains the media description offer generated by the SRVCC enhanced MSC Server and MGW reserved for the given session transfer call.
  - The c-attribute contains the IP address of MGW.
  - The m-attribute contains the IP port number of RTP media of MGW.

- The individual speech codec(s) are listed in the priority order created based on the supported codec list information that was received from MME in the PS-CS handover request.

The SRVCC enhanced MSC server having an integrated MGCF will receive an acknowledgement eventually from the serving IMS domain with the 200 OK (invite) message.

In the case where the SRVCC enhanced MSC server would not have an integrated MGCF then it is also possible to use a traditional circuit switched network network–network interface (NNI) protocol such as ISDN user part (ISUP) or bearer independent call control (BICC) to the route session transfer call via a separate MGCF to IMS/SCC AS. This issue depends on the capabilities of the equipment used.

It should be highlighted here that the complete SRVCC procedure consists of preparation and actual domain transfer phase. It is possible that the preparation phase happens and the required bearer resources for domain transfer towards target CS network are reserved but the actual domain transfer phase is not performed afterwards because for instance the served UE has moved again back to a good LTE coverage or eNb cancels the HO request. However in the case where the domain transfer phase has been started and SCC AS is contacted by an SRVCC enhanced MSC server then UE is moved to CS radio access. Reverse SRVCC is needed to perform domain transfer from CS to LTE radio access (this feature is being specified as part of 3GPP Release 11).

Finally when session transfer call establishment has been completed by an SRVCC enhanced MSC server towards SCC AS and no cancellation has been received from EPS then the SRVCC enhanced MSC server will acknowledge MME that the required procedures has been completed by sending the SRVCC PS-CS complete notification message to MME and prepare to receive a handover complete indication from target RNC or BSC (possibly via other MSC-B/MSC server-B) that UE has successfully moved to coverage of UTRAN or GERAN.

The SRVCC PS-CS complete notification message contains the information listed in Table 5.8.

Table 5.8 SRVCC PS-CS complete notification message

Name of parameter	Purpose	Example value/content
IMSI	This parameter identifies the subscriber performing the domain transfer. SRVCC enhanced MSC server uses this towards MME to identify subscriber in question.	Identifier of subscription (15 digits long) including mobile country code, mobile network code and actual mobile subscriber identity number (MSIN).
Private extension	This parameter is optional and can be used to transfer vendor specific information between network elements.	Any

After handover complete has been received by the SRVCC enhanced MSC Server it may perform re-allocation of a new temporary mobile subscriber identity (TMSI) for UE via the circuit switched radio access network and it will perform the location update procedure towards HSS in order to inform HSS that the terminating circuit switched calls and short messages have to be sent to the subscriber via the SRVCC enhanced MSC server. It should be noted that when the subscriber's (Kristiina in this example) UE has been also attached into the LTE via a combined EPS/IMSI attach procedure, that is has been able to use CS fallback for EPS then the location update procedure performed by the SRVCC enhanced MSC server will replace the VLR address of that particular MSC server serving the SGs association for that given subscriber.

Figure 5.25 represents how call control signalling, user plane (speech) connection as well as Internet-based applications are connected after domain transfer has been successfully completed.

After domain transfer, the dashed line illustrates the call control signalling connection via a circuit switched domain to the serving IMS domain (SCC AS). In this particular example the signalling connection is routed via the MSC-B (MSC server) controlling the circuit switched radio access network (3G) via an SRVCC enhanced MSC server and separate MGCF to the SCC AS located in the serving IMS domain. As described above the SCC AS has performed the inter-domain mobility with minimum impact to the remote party. The solid line illustrates the user plane (speech) connection that is now routed via circuit switched access (MGWs) to the remote side of call. Finally also the IP traffic that continues towards multiple active Internet-based applications after domain transfer via UTRAN radio access and packet domain is shown with a dotted line from UE via Nb, RNC, SGSN and S/P GW to the Internet.

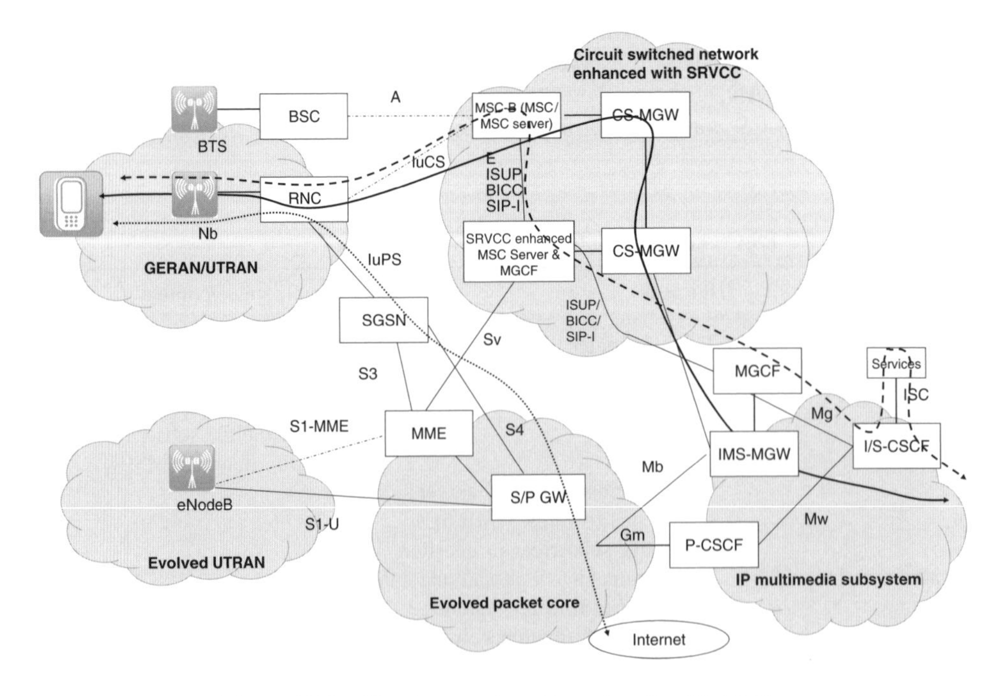

Figure 5.25 Signalling and media paths after SR-VCC.

As can be seen from the example, the domain transfer procedure for an ongoing VoLTE call is rather complex one. Also the performance of this procedure depends on the way in which the involved network elements and UE are able to work together in the most optimal way to reduce the impact for ongoing call connection. The requirement for maximum speech connection gap is defined in (3GPP TS 25.913) to be 300 ms. Above this it is expected that end user experience will be degraded.

5.6.2.2 Evolution of SRVCC

Since the magnitude of SRVCC required standardisation as well as a number of requirements it was divided into phases inside the 3GPP standardisation process.

The previously described basic functionality was completed within 3GPP Release 8 but then other enhancements were introduced into 3GPP Releases 9, 10 and 11.

3GPP Release 9 introduces a specific 'MSC server assisted mid-call' functionality in (3GPP TS 23.237) that is derived from the ICS enhanced MSC server functionality defined in (3GPP TS 23.292). This enhancement introduces a new SIP interface (i2) into the SRVCC enhanced MSC server, which makes it possible to exchange session state information directly between SCC AS and SRVCC enhanced MSC server, which is not possible when, for example ISUP is used. This session information transfers the state of the ongoing sessions as known by the SCC AS to the SRVCC enhanced MSC server, making it possible that the SRVCC enhanced MSC server is then aware that the end user had two ongoing sessions (one held and one active). Additionally the use of i2 enables control of the conference session that was previously established in IMS by the served UE (i.e. before SRVCC occurred).

Table 5.9 represents the end to end functionalities that are enabled by different 3GPP releases within the scope of domain transfer procedure (SRVCC).

Figure 5.26 represents the 3GPP Release 10 level network architecture for domain transfer constituting changes from previous releases. This architecture is shown in order to achieve a better understanding about the differences between the 3GPP Release 8 and 10 architectures. 3GPP Release 8 architecture was shown earlier in this section.

The GSMA VoLTE (IR.92) profile mandates the Release 8 baseline, which means that most of the networks in the early phase will likely be compliant with that release. However it is expected that networks will be enhanced to support a later 3GPP baseline eventually via normal network evolution.

5.6.3 Summary

This section described an example of mobility management scenarios, which may be applied for a VoLTE session when the UE moves between different kinds of 3GPP radio access network technologies.

The first part described mobility within two packet switched domains (3G/UTRAN and LTE/E-UTRAN), which can also be supported in the case when the legacy radio access technology for 3G has been upgraded to support voice over IP with conversational traffic class, robust header compression and other relevant technologies.

The second part described mobility between a packet switched domain (LTE/E-UTRAN) and a circuit switched domain. This mobility enables the operator to deploy

Table 5.9 SRVCC capabilities in 3GPP releases

3GPP release	End to end capability	Notes
Release 8	Domain transfer is supported only for single active voice session.	This is currently selected as the baseline for the GSMA VoLTE profile and therefore it is expected that Release 8 will be supported by all VoLTE networks at minimum.
	Held sessions or sessions in the alerting phase are cleared and only an active one is transferred to the circuit switched domain.	
Release 9	In addition to the previous release also:	In order to support these capabilities enhancement for SRVCC an enhanced MSC server is needed. This enhancement is called the mid-call assisted MSC server (MaM), as described earlier.
	Held sessions and IMS based ad hoc conference session are transferred.	In addition to MaM also a new functionality is introduced into the network architecture, called the emergency access transfer function (EATF) that is used to anchor the emergency session and which then receives the session transfer call from the SRVCC enhanced MSC server by using the emergency session transfer number for SRVCC (E-STN-SR).
	Emergency sessions can be transferred.	
Release 10	This release optimises the domain transfer (SRVCC) network architecture in order to reduce any possible speech path gap in case of roaming scenarios.	New functionalities are introduced into the network architecture, called the access transfer control function (ATCF) and access transfer gateway (ATGW) which are both located in the serving network.
	Anchoring of the session is performed always in the serving network in order to achieve this.	ATCF is responsible to handle the anchoring of the SIP session and ATGW anchors the speech connection already at the beginning of a session (RTP/RTCP).
	In addition, will also introduce support for alerting phase access domain transfer.	SRVCC enhanced MSC server will establish a session transfer call directly to the ATCF, which again will control the ATGW to switch the speech path towards a MGW reserved by the SRVCC enhanced MSC server. This way another side of the connection does not have to be updated, as in the previous example.

Table 5.9 *(continued)*

3GPP release	End to end capability	Notes
Release 11	Support for reverse SRVCC from GERAN/UTRAN to E-UTRAN.	Also CS originated calls from a VoLTE capable UE require to be routed via IMS in order to let SCC AS have knowledge of these calls for a possible reverse SR-VCC.
	SRVCC for video calls.	SRVCC enhanced MSC server is also able to support domain transfer for CS video telephony, according to 3GPP 3G-324M to LTE.
		No major architectural improvements beyond Release 10.

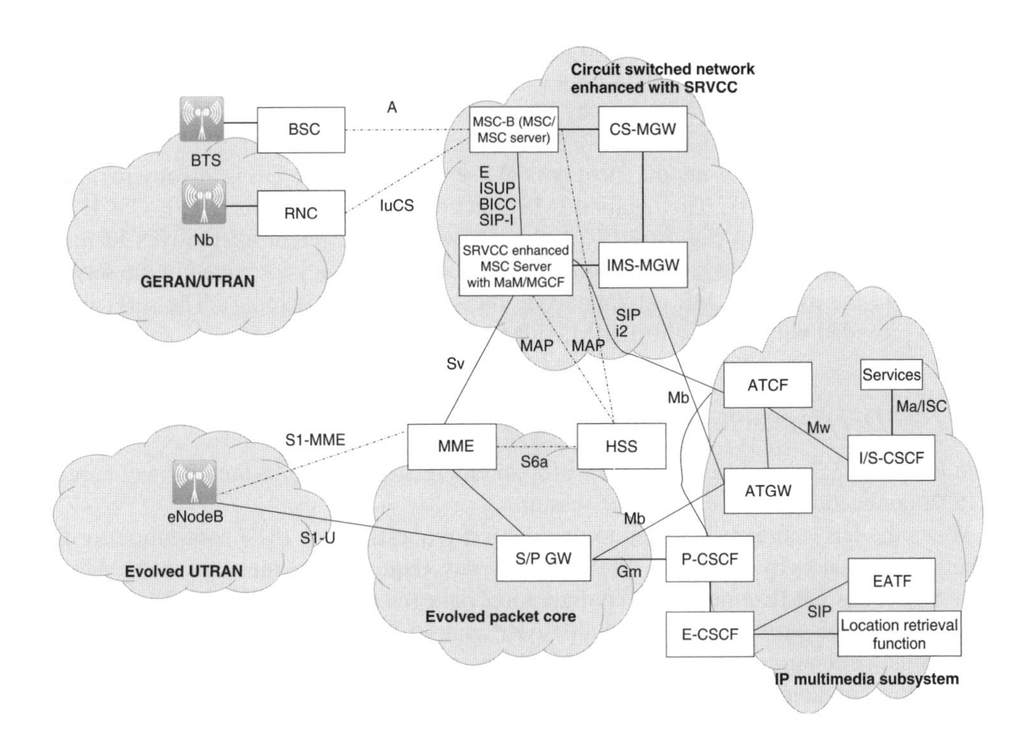

Figure 5.26 SR-VCC 3GPP Release 10 architecture.

LTE by having a minimum impact to the other radio access technologies in place. However in this case, due the complexity of domain transfer procedure (SRVCC), it may be that the initial VoLTE deployments are made without this feature. A target standardisation baseline for domain transfer is expected to be 3GPP Release 10 but initial deployments may be done already using the 3GPP Release 8 baseline.

5.7 IMS Emergency Session

In the case of an emergency session special procedures are required to ensure that the user receives special treatment in access, EPC and IMS domains.[12] The following procedures are performed during the IMS emergency session after the UE has detected an emergency session request from the user (see also Figure 5.27):

- UE conducts RRC connection establishment with an emergency indication.
- UE requests PDN connectivity with an emergency indication.
- MME selects an emergency APN based on emergency configuration data and requests the PDN GW to establish a default bearer. This is done in parallel to an existing PDN connectivity to IMS APN.
- UE executes an IMS emergency registration using the received IP address and P-CSCF address. This registration is done in parallel to an existing regular registration.
- UE sends an emergency session request to IMS.
- IMS routes the emergency session to a public safety answering point (PSAP) and delivers the session information to PCRF.
- The PCRF creates and delivers a policy rule to P-GW.
- P-GW sets up a dedicated emergency bearer.

The description here covers the most typical use case where the UE is already regularly registered to the network. The detailed IMS function configuration such as E-CSCF and LRF can be found from Figure 3.17 (it also shows three different alternatives to reach PSAP). The emergency service is not a subscribed service and it is provided in the serving network (in the roaming IMS network while the subscriber is roaming). The service can be also provided without a valid USIM, ISIM or roaming agreement.

5.7.1 PDN Connection Setup for Emergency Session

The UE may make an initial attachment for an emergency session or the UE may request a PDN connection for an emergency session.

When the UE is already attached to the MME it requests a new PDN connection for the emergency session by sending a PDN connectivity request NAS message to the MME. The MME has MME emergency configuration data (emergency APN, P-GW ID, APN AMBR and QoS profile, which includes ARP and QCI). The MME emergency configuration data contain the emergency APN which is used to derive a P-GW, or the MME emergency configuration data may also contain the statically configured P-GW for the emergency APN. The MME continues the procedure by sending a create session request to S/P-GW.

The MME sends bearer setup request to the eNB including the ARP values for emergency bearer services indicate the usage for emergency services to the E-UTRAN.

[12] In Finland 3 954 569 emergency calls were made in 2010 (http://www.112.fi/documents/Tilastoja_tammi-joulu_2011.pdf). This means 0,74 emergency calls per inhabitant annually.

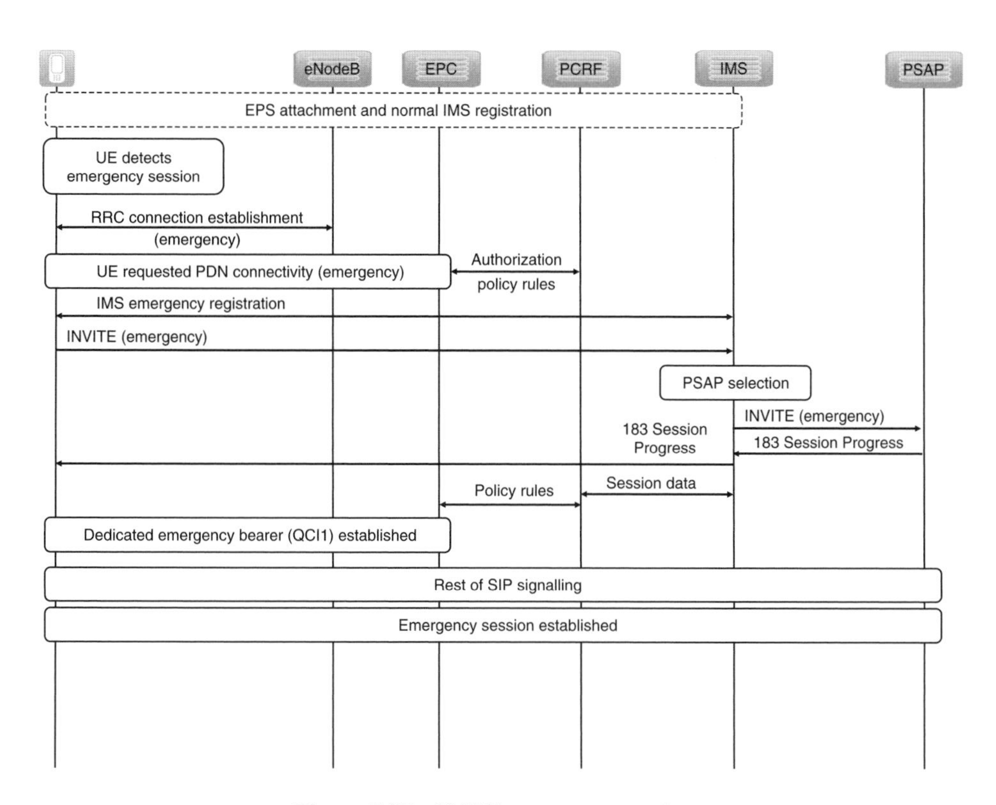

Figure 5.27 VoLTE emergency session.

After radio bearer setup the UE can use an emergency PDN connection for the SIP signalling.

5.7.2 *Emergency Registration*

The UE has to perform the IMS emergency registration when the UE has a valid USIM/ISIM and the EPC has indicated it supports emergency specific functionalities as part of EPS attachment or as part of tracking area updates and UE has detected that the subscriber has dialled one of common emergency numbers, such as 911, 112.

The emergency registration procedure follows normal IMS registration procedures, as defined in Section 5.4 with the exceptions described here.

The REGISTER request constructed by the UE contains a special ‘SOS’ parameter in the Contact header to indicate that this registration is for emergency purposes and the Contact header does not contain feature tags for MMTel and SMS over IP. The emergency REGISTER request using the new IP address would look like this:

```
REGISTER sip:ims.mnc005.mcc244.3gppnetwork.org SIP/2.0
Via: SIP/2.0/UDP [5555::1:1:1:1]:1600; branch=0uete
Max-Forwards: 70
```

```
P-Access-Network-Info: 3GPP-E-UTRAN-TDD; utran-cell-id-3gpp=
     244051F3F5F7
From: <sip:kristiina@example.com>;tag=3442f
To: <sip:kristiina@example.com >
Contact: <sip: [5555::1:1:1:1]:1600>;
expires=600000;
sos;
+sip.instance="<urn:gsma:imei:90420156-025763-0>";
Call-ID: herhe4636hdhfh3466
Authorization: Digest username="private_user1@example.com",
        realm="ims.mnc005.mcc244.3gppnetwork.org", nonce="",
        uri="sip:ims.mnc005.mcc244.3gppnetwork.org",response=""
Security-Client: ipsec-3gpp; alg=hmac-sha-1-96; spi-c=1111;
                             spi-s:=2222; port-c=9999; port-s=1600
Require: sec-agree
Proxy-Require: sec-agree
Supported: path
CSeq: 10 REGISTER
Content-Length: 0
Content-Length: 0
```

This request gets routed to S-CSCF, and the request is challenged, and the UE responds to the challenge as described in Section 5.4. When the S-CSCF receives the second REGISTER containing a valid challenge response it will accept the registration but according to local regulation it sets the registration timer value (Expire header) to an emergency session specific value. This value is expected to be much shorter than normal registration. The emergency registration should be active for the duration of emergency session and for some time after session completion to allow PSAP to call back to the caller. Furthermore, the S-CSCF is not expected to set a Service-Route header as the emergency session is routed from the P-CSCF to an emergency-call session control function (E-CSCF) not to the S-CSCF.[13] The 200 OK response would look like this:

```
SIP/2.0 200 OK
Via: SIP/2.0/UDP icscf1.example.com;branch=3icte
Via: SIP/2.0/UDP pcscf1.example.com;branch=2pcte
Via: SIP/2.0/UDP [5555::1:1:1:1]:1600;branch=1uete
From: <sip:kristiina@example.com>;tag=89797
To: <sip:kristiina@example.com>;tag=sgdewe3
P-Associated-URI: <sip:kristiina@example.com>, <tel:+358501234567>
P-Charging-Function-Addresses: ccf=chargingentity.example.com
P-Charging-Vector: orig-ioi: "Type1  example.com"; term-ioi:
     "Type 1 example.com"
Path: <sip:term@pcscf1.home.fi;lr>
Contact: <sip: [5555::1:1:1:1]:1600 >;
            expires=1000;
            sos;
            +sip.instance="<urn:gsma:imei:90420156-025763-0>";
Call-ID: herhe4636hdhfh3466
```

[13] If a communication service provider desires to route a non roaming emergency session via S-CSCF then the S-CSCF adds a Service-Route header.

```
CSeq: 11 REGISTER
Content-Length: 0
```

In contrast to a normal registration UE and P-CSCF do not subscribe to registration event package (Section 5.4.10). Furthermore, the UE or network are not allowed to execute deregistration (Section 5.4.12). In other words the emergency registration stays active until the registration timer expires (here 1000 s). UE can, however, perform user-initiated emergency re-registration if the registration timer is about to expire when the UE is still engaged on an emergency communication or is in the process of initiating an emergency session.

5.7.3 Emergency Session

After the registration procedure the UE constructs an emergency INVITE request (required SDP information is not shown here; see Section 5.5.1 for SDP description).

```
INVITE urn:service:sos SIP/2.0
Via: SIP/2.0/UDP [5555::1:1:1:1]:1600; branch=abc123
Max-Forwards:70
Route: <sip:[5555::55:66:77:88]:9000;lr>
P-Access-Network-Info:3GPP-E-UTRAN-TDD;utran-cell-id-
     3gpp=2440051F3F5F7
P-Preferred-Identity: <tel:+358501234567>
P-Preferred-Identity: <sip:kristiina@example.com>
Privacy: none
From: <sip:kristiina@example.com>;tag=171828
To: urn:service:sos
Call-ID: cb03a0s09a2sdfglkj490333
Cseq: 127 INVITE
Require: sec-agree
Proxy-Require: sec-agree
Supported: precondition, 100rel, 199
Security-Verify: ipsec-3gpp; alg=hmac-sha-1-96; spi-c=98765432;
     spi-s=87654321; port-c=8642; port-s=9000
Contact: <sip:5555::1:1:1:1]:1600;
Allow: INVITE, ACK, CANCEL, BYE, PRACK, UPDATE, REFER, MESSAGE,
     OPTIONS
Accept:application/sdp, application/3gpp-ims+xml
Content-Type: application/sdp
Content-Length: (. . .)
```

This request looks pretty much like a normal INVITE request (Section 5.5.1). The differences here are:

- Request-URI and To headers contain the emergency service uniform resource name (URN), which UE has translated from the dialled number (e.g. 112 or 911).
- The P-Preferred-Identity headers contain Kristiina's registered IMS public user identities (one SIP URI and one Tel URI).[14]

[14] The P-Preferred-Identity header is mandatory for emergency sessions whereas it is optional for normal sessions.

- The Route header contains only address of P-CSCF.
- Information related to the MMTel service are not present (the Accept-Contact and P-Preferred-Service headers are not present and the Contact header does not contain MMTel ICSI).

When this request arrives at the P-CSCF it will select an E-CSCF within the same IMS network and routes the request to it. The E-CSCF will utilise the UE provided location information (as transported in the P-Access-Network-Info header) and the UE provided emergency type (if provided, like sos.fire, sos.police) to resolve the PSAP address. When more accurate location- or network-provided information is required then the E-CSCF will use the LRF function (see Section 3.4.4.4) to obtain location information or to select the actual PSAP. The request is further routed to the PSAP and here on the session setup follows the principles of a normal session setup.

If PSAP requires more accurate or updated location information from the caller it will send a location request to the location retrieval function (LRF). The LRF will either use a secure user plane location (SUPL) standard defined by the Open Mobile Alliance or control plane solution defined by 3GPP to obtain position information.

5.7.3.1 Related Standards

- 3GPP TS 23.167 – IP multimedia subsystem (IMS) emergency sessions.
- 3GPP TS 23.401 – General packet radio service (GPRS) enhancements for evolved universal terrestrial radio access network (E-UTRAN) access.
- 3GPP TS 24.229 – IP multimedia call control protocol based on session initiation protocol (SIP) and session description protocol (SDP); stage 3.
- 3GPP TS 29.212 – Policy and charging control over Gx reference point.
- 3GPP TS 29.214 – Policy and charging control over Rx reference point.
- RFC5031 – A uniform resource name (URN) for services.

5.8 CS Fallback for Evolved Packet System Call Case(s)

CS Fallback for EPS (CSFB) introduces capability to re-use existing circuit switched core network services as such while still enabling terminal to camp in LTE radio access. CSFB has been initially standardised in 3GPP Release 8 content and further enhanced in 3GPP Release 9 with number of improvements that are expected to increase end user experience. The stage 2 specification defining the systemwide functionality of CSFB is (3GPP TS 23.272).

At a high level CSFB has been defined into multiple functionalities (use cases) that can be deployed separately based on the market need. These functionalities include the following use cases as defined in the described 3GPP standardisation releases (in brackets):

- Mobile originating SMS over LTE (3GPP Release 8);
- Mobile terminating SMS over LTE (3GPP Release 8);
- Mobile originating circuit switched voice and video telephony call (3GPP Release 8);
- Mobile originating emergency call (3GPP Release 8);
- Mobile originated priority call (3GPP Release 10);

- Mobile terminating circuit switched voice and video telephony call (3GPP Release 8);
- Mobile originating call independent supplementary service procedure (3GPP Release 8);
- Network initiated unstructured supplementary service data (NI USSD) (3GPP Release 8);
- Mobile terminating location request (MT-LR) (3GPP Release 8).

It is assumed that the initial deployments of CSFB are focusing on data-centric devices (probably tablets, USB dongles or similar form factors) that do not have integrated voice telephony. Therefore initial use cases were centered on SMS and especially on how a short message can be transferred from network to terminal for instance in order to configure actual terminal settings (device management) or deliver information about cost of data service (roaming bill shock prevention).

Both mobile originated SMS and mobile terminating SMS capability were part of the use cases. In the long run native SMS over IP as defined by 3GPP, which is described in Section 5.9.2.2, is expected to be taken into use with UEs that also support native IMS-based VoIP, but support for SMS over LTE with CSFB is also expected to be used in parallel to this for a long time.

In order to support voice services in manner similar to that supported by mobile networks today, it is assumed that user cases around mobile originating and terminating voice and video telephony calls are made possible. This means that voice-centric devices having an integrated telephony client (probably based on the handset form factor) which camp in LTE radio access are able to either originate or terminate calls by performing a fallback to legacy radio access network which can be GERAN, UTRAN or CDMA2000 1x, in order to complete the actual call. In fact such UE having voice-centric nature will not even camp on LTE unless the core network does not provide suitable voice capable service for instance via CSFB. This means that it is expected that most networks will eventually implement CSFB in order to successfully deploy voice service for LTE attached UEs. Therefore the speech path involved in an actual established call is always made via legacy radio access technology rather than via LTE. After the call has been completed then UE moves back to coverage of LTE, if under sufficient coverage, or continues camping in the 2G/3G cell. This is important in order to re-enable the benefits of LTE when possible.

The remaining call independent user cases, including location retrieval, may also be required in some practical deployments. Especially USSD has many applications that continue to maintain importance unless applications are re-implemented by using native IP technology. Such applications are for instance related to the use of pre-paid services.

In the long run the role of CSFB may gradually diminish in the network when VoLTE based on IMS will be introduced and penetration of IMS capable terminals increase. The introduction of IMS-based VoIP may be done in a phased manner, meaning that in the first phase IMS-based VoIP is supported only when subscriber is within home public land mobile network (PLMN) but when subscriber roams to other LTE network then pure CS voice or CSFB will be used instead when native IMS roaming architecture is not yet in place. Additionally in the case where the serving network does not support an IMS-based emergency call or priority service as required by the local regulator then the operator may use CSFB for these services instead of IMS at beginning. This enables gradual introduction of IMS-based VoIP depending on the business and technological capabilities of network and UEs. The expected target architecture will be that emergency call, multimedia priority service as well as roaming will be used based on IMS.

The remaining part of this subsection describes each user case in a more detailed fashion as well as describing the way that CSFB can be architecturally implemented into a network. Additionally the main focus is on a 3GPP-based architecture and solution whereas it is expected that CSFB will have no significant role in 3GPP2 markets.

5.8.1 Architecture of CS Fallback for EPS

In order to implement CSFB capability to LTE deployment changes are required to both the existing circuit switched network as well as the EPS. Possibly the most visible change is related to the introduction of a so-called SGs interface as defined in (3GPP TS 29.118) between MME serving LTE users capable of CSFB functionality and the MSC or MSC server of circuit switched networks. However the magnitude of changes required to circuit switched network depends on the selected approach, that is whether an overlay deployment is used or not.

The CSFB functionality can be used in the case where UE, E-UTRAN and relevant core network domains support the functionality. This means that there is no specified subscription-level provisioning needed for CSFB in HSS/HLR but MME can reject a request from the UE to perform combined IMSI/EPS attachment, if so configured. In case EPS rejects combined IMSI/EPS attachment then in case of voice-centric LTE smartphone, when no IMS based voice service is available, it usually means that UE will camp on legacy 3G/2G or CDMA technology.

Figure 5.28 represents a generic 3GPP network providing CSFB.

In the architecture shown (Figure 5.28) the SGs interface is deployed to all such MSC/MSC server network elements that provide GERAN/UTRAN service for same geographical area rather than the E-UTRAN served by EPS. Additionally it is possible to deploy this functionality into MSC/MSC server network elements which are pooled as

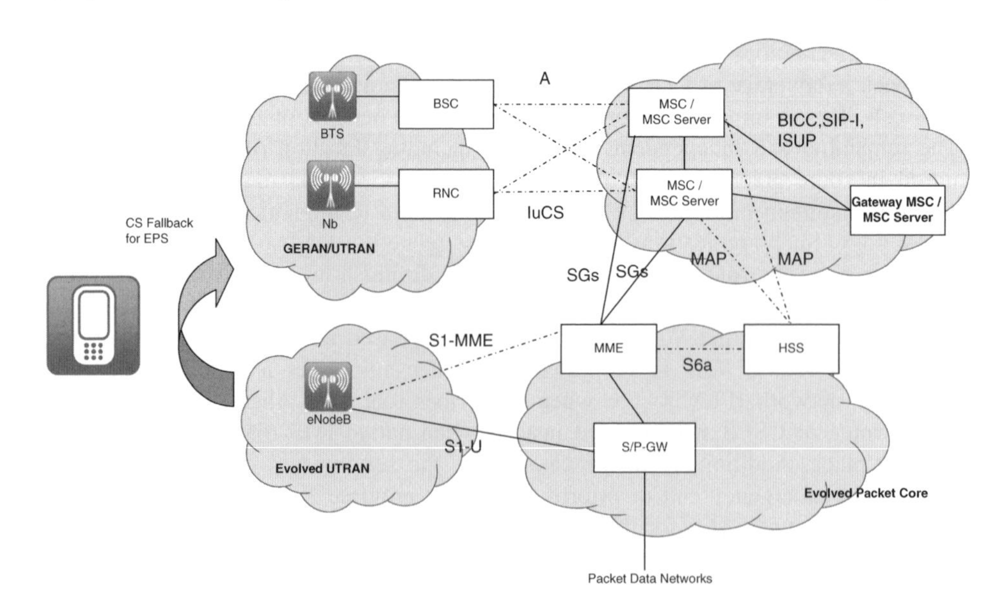

Figure 5.28 Architecture for CS fallback.

described in (3GPP TS 23.236). In that case pooling mechanisms [the network resource identifier (NRI) part of TMSI] can be used to route signalling messages correctly to the right MSC/MSC server element in the pool after fallback has been performed. This kind of network architectural deployment is described in Figure 5.29.

As the configuration of SGs interface is the point to point connection between MMEs and MSC/MSC servers, it means that it is possible and fairly likely to design a network in such a manner that a single MME is connected to multiple MSC/MSC servers and vice versa. In the case where a single MME is connected to multiple MSC/MSC servers the selection of MSC/MSC server can be based on mapping between the tracking areas served by MME, old LAI received from UE and the location areas served by the MSC/MSC server. It can be assumed that by default CSFB functionality for UEs within a single tracking area are handled by at least a single MSC/MSC server. However in the case where an increased network level resiliency is required then it is allowed by (3GPP TS 23.272) that MME performs mapping of the location area from the tracking area in such a way that multiple MSC/MSC servers are selected from a 'pool' of a configured list of MSC/MSC servers. MME should follow principles defined in (3GPP TS 23.236), which is also the basis for traditional circuit switched and packet switched pool functionality. According to this specification a specific IMSI hash functionality is performed for each combined IMSI/EPS attached UE during the attachment phase that results in the identity of the MSC/MSC server allocated to serve a particular UE.

Additionally the very first deployment of CSFB may be done in order to support only SMS over LTE (SGs), for instance due to the reason that only data-centric devices are available commercially. In this case if no voice call or other use cases are required, then the network configuration in which MME is connected only to a few MSC/MSC Server in network can be used without placing additional requirements to other network entities. This means that data-centric devices performing 'SMS-only' combined IMSI/EPS attached to EPS can be easily served by a few overlay MSC/MSC servers since no actual fallback is performed from one E-UTRAN to GERAN/UTRAN/CDMA2000 1x radio access.

However in the case where other user cases of CSFB are required, especially a mobile terminating call, then this kind of network deployment requires the use of a so-called mobile terminating roaming retry (MTRR) functionality that was originally standardised as part of (3GPP TS 23.018) and endorsed in (3GPP TS 23.272). This MTRR procedure was originally standardised in order to solve problems caused by legacy circuit switched mobile phones that hopped between the 2G and 3G radio access technology, causing terminating calls to fail (or forward to voice mail system) due to an unreachable condition. Re-use of MTRR in scope of CSFB requires that HLR and gateway MSC/MSC servers in home PLMN support this functionality. The same applies also to those networks from where inbound roamers can roam to particular EPS and use CSFB. The MTRR procedure is described in more detail in Section 5.8.9.1.

Eventually the higher complexities involved in the deployment and use of MTRR (including end user experience penalties due to longer call establishment time as well as the impact caused for network elements located in home PLMN) led to such a decision that the new procedure was proposed to be standardised as part of 3GPP Release 10, named mobile terminating roaming forwarding (MTRF). This procedure causes changes mainly to serving PLMN and removes the need to re-route calls from the home PLMN in the case where the serving MSC or MSC server changed during fallback procedure. A more detailed description of this procedure is given in Section 5.8.9.2.

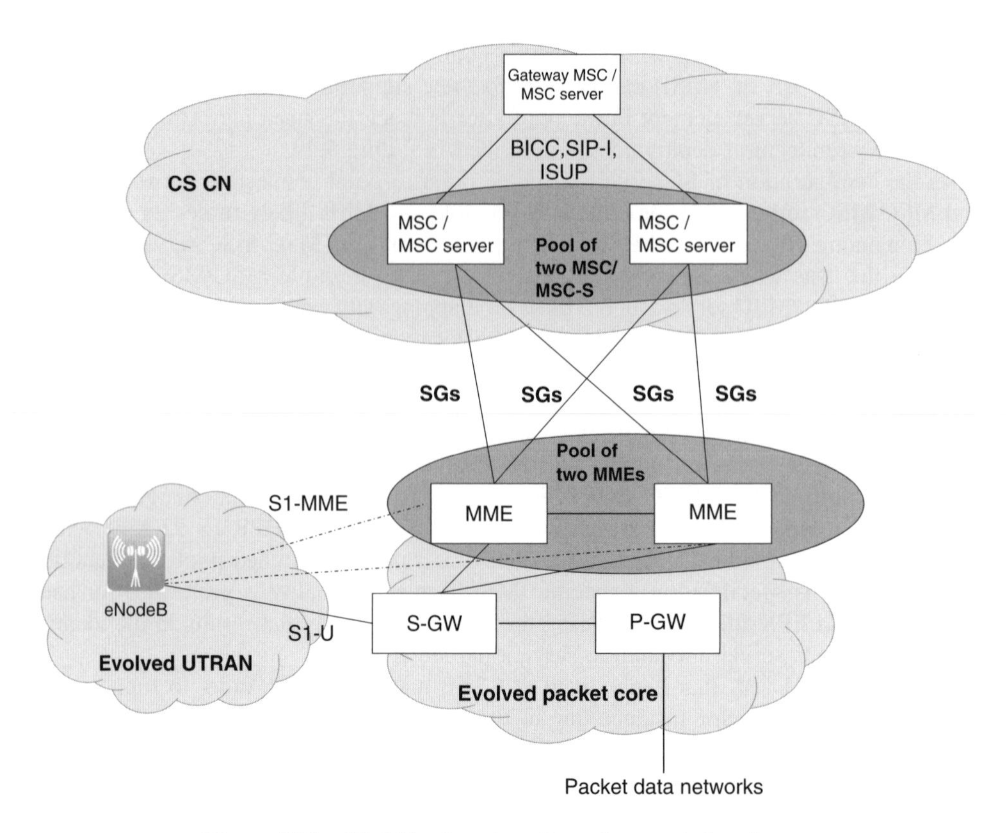

Figure 5.29 CS fallback and pooling of network functions.

Figure 5.29 represents this network configuration which provides increased resiliency against MSC/MSC server network element failures.

Finally, as also shown in Figure 5.29, it is also likely that MMEs are configured into a 'pool' of MMEs that provide a service for a single tracking area in order to increase the resiliency of EPS. This is so called S1-flex (pool).

5.8.2 Description of SGs Interface

The SGs interface defined in (3GPP TS 29.118) was standardised based on high level principles of the Gs interface as specified in (3GPP TS 29.018) used today in many mobile networks to provide paging co-ordination for terminals capable for both PS data and CS voice.

Despite the fact that underlying principles have been taken from Gs interface the SGs interface is based on different design principles than Gs. SGs interface is first of all based on a native stream control transfer protocol (SCTP) without a classical SS7 protocol layer [MTP3/signalling connection control part (SCCP)] whereas the Gs interface is based on SS7 routing principles. This means that the SGs interface is always directly established between the MME and the involved MSC/MSC server.

Figure 5.30 represents the protocol stack of the SGs reference point.

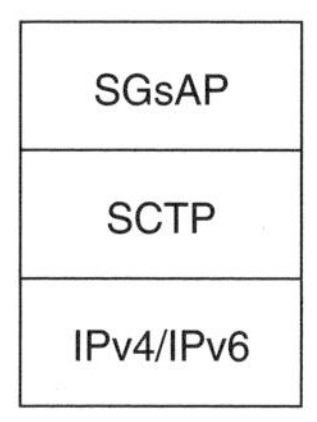

Figure 5.30 Protocol stack of SGs reference points.

SGsAP represents the SGs application part that is the actual protocol payload for the SGs interface including message syntax and information element structure. SGsAP is transported over the SCTP which can support either multihoming for path protection against backbone failures or only single homing functionality, depending on the capabilities available in the actual network elements. These SCTP associations are pre-configured and taken into use via management actions performed to network the elements in question. The use of multihoming is a widely recommended functionality in order to provide increased network level resiliency against IP connectivity failures. Finally both IPv4 and IPv6 are supported by the SGs interface but again deployment can be done in flexible manner depending on the capabilities of the actual network elements.

The SGs interface message syntax is based on traditional tag length value (TLV) encoding containing the actual message definition and well as mandatory and optional information elements. Similarly the information element contains the same kind of TLV structure.

SGsAP related messages are used by certain standardised procedures that are performed between MME and MSC/MSC server. Particular procedures may consist of exchange of only single SGsAP message or alternatively it consist sequence of multiple SGsAP messages. SGs protocol defines, for example the following procedures:

- Alert and UE activity indication;
- EPS detach and IMSI detach;
- Downlink unitdata and uplink unitdata;
- Location update;
- Paging;
- MM information;
- Reset and status.

Actual detailed encoding of these all these previously listed procedures is beyond the scope of this book. Relevant procedures are referenced in a later part of this chapter when actual call flows are described. Additionally the interested reader is advised to investigate (3GPP TS 29.118) to get more familiar with actual protocol level details of the SGs interface, as shown in Table 5.10.

5.8.3 Idle Mode Signalling Reduction and Use of CS Fallback for EPS

In early LTE deployments due to spotty LTE coverage it may be required that the idle mode signalling reduction (ISR) feature in LTE is used. This functionality makes it possible

Table 5.10 SGsAP messages

SGsAP message	Purpose
SGsAP-alert-ack	Used by MME to acknowledge initiation of alert procedure towards the MSC/MSC server
SGsAP-alert-reject	Used by MME to reject initiation of alert procedure towards the MSC/MSC server
SGsAP-alert-request	Used by the MSC/MSC server to request an alert procedure from MME An alert request is needed since the mobility state of an EPS attached UE is logically divided between MME and MSC/MSC server. It means that the MSC/MSC server (VLR) does not have always up to date knowledge of the reachability of the EPS attached UE via an SGs interface and thus VLR need to request MME to inform itself regarding this state.
SGsAP-downlink-unitdata	Used by the MSC/MSC server to transfer transparently downlink short message service-related information (e.g. actual content of short message) towards the EPS.
SGsAP-EPS-detach-ack	Used by the MSC/MSC server to acknowledge a successful detachment of UE from EPS either by MME or UE. After this procedure the UE is no longer considered to have an SGs association.
SGsAP-EPS-detach-indication	Used by MME to initiate EPS detach procedure towards the MSC/MSC server in order to close the SGs association established prior this event.
SGsAP-IMSI-detach-ack	Used by MSC/MSC server to acknowledge a successful detachment of the IMSI (UE) from EPS either by MME or UE. After this procedure the IMSI (UE) is no longer considered to have an SGs association.
SGsAP-IMSI-detach-indication	Used by MME to initiate the IMSI detach procedure towards the MSC/MSC server in order to close an SGs association established prior this event.
SGsAP-location-update-accept	Used by the MSC/MSC server to acknowledge a successful location update of a particular IMSI to VLR by MME. This procedure is triggered by a combined IMSI/EPS attach procedure and establishes an SGs association. After this procedure the IMSI (UE) has an SGs association.
SGsAP-location-update-reject	Used by the MSC/MSC server to reject a location update procedure towards MME. The outcome of this is that the IMSI (UE) is not considered to have successfully established an SGs association.
SGsAP-location-update-request	Used by MME to initiate a location update procedure towards the MSC/MSC server. Successful execution of this procedure is a pre-requisite for any use of CSFB functionality. It should be noted that an SGs interface does not use a periodical location update procedure similar to the A/IuCS interface.

Table 5.10 *(continued)*

SGsAP message	Purpose
SGsAP-MM-information-request	Used by the MSC/MSC server to transfer mobility management layer information towards a UE attached into EPS. This includes for instance network information, time zone (NITZ) information in order to synchronise the time of date for the UE from the core network.
SGsAP-paging-reject	Used by MME to reject a paging request from the MSC/MSC server. The outcome of this is that a terminating call, SMS, LCS or USSD towards an EPS attached UE cannot be completed.
SGsAP-paging-request	Used by the MSC/MSC server to request paging of a UE for a terminating call, SMS, LCS or USSD towards a UE attached into EPS. The UE will be informed about this request by MME and in the case where RRC connection does not yet exist (the UE is in EPS idle-mode) then such a connection is established by MME.
SGsAP-reset-ack	Used by MME or the MSC/MSC server to acknowledge a successful outcome of a reset procedure initiated by another network element.
SGsAP-reset-indication	Used by MME or the MSC/MSC server to inform the other side of an SGs interface that a particular network element has been restarted and the SGs association data has been lost. This will cause UEs to be unreachable for any terminating requests (paging) until the SGs associations are again established by performing a location update procedure. The trigger for such a location update may be for instance a periodical tracking area update performed by an EPS attached UE.
SGsAP-service-request	Used by MME towards the MSC/MSC server as a response to a paging request via SGs in order in order to inform that MME has successfully established a radio signalling connection towards UE and a terminating request can proceed. This message resembles an acknowledgement to a paging request even though named differently.
SGsAP-status	Used either by MME or the MSC/MSC server as a result of a recognised protocol syntax error from any received message, which can be for instance: Unknown message which is not recognised by recipient but ignored; Missing mandatory information element: Syntactically or semantically incorrect information element received.
SGsAP-TMSI-reallocation-complete	Used by MME to indicate towards the MSC/MSC server of a successfully performed TMSI reallocation via an SGs interface.

(continued overleaf)

Table 5.10 (*continued*)

SGsAP message	Purpose
SGsAP-UE-activity-indication	Used by MME to indicate towards the MSC/MSC server that an EPS attached UE has performed some activity within reallocation via an SGs interface and is reachable for instance to receive a short message.
SGsAP-UE-unreachable	Used by MME to indicate towards the MSC/MSC server that EPS has not been able to reach a targeted UE as the result of a paging request via SGs. This procedure may result in the activation of an alert procedure by the MSC/MSC server in order to deliver succeeding activity indications from MME to MSC/MSC server.
SGsAP-uplink-unitdata	Used by MME to transfer transparently uplink short message service related information (e.g. actual content of short message) towards the MSC/MSC server.
SGsAP-release-request	Used by the MSC/MSC server (VLR) towards MME in order to inform MME that no more network access stratum (NAS) messages are assumed to be transferred between VLR and UE.

that UE moves in idle-mode between different 3GPP radio access networks (2G/3G/LTE) having concurrent registration to both the serving SGSN and MME network elements. On the HSS side this is seen as a reduced amount of signalling procedures since mobility between 3GPP-based radio technologies no longer causes as much signalling towards HSS as would be required in the case where ISR is not used at network level.

Use of ISR requires that UE, serving 2G/3G and LTE network as well as HSS support it. The network may enable or disable the use of ISR if required.

The downside in the use of ISR is related to negative side-effects that will be caused by it for VoLTE and CSFB. In the case of CSFB it means that if ISR is supported for a given UE in idle-mode having a combined IMSI/EPS attachment to LTE for CSFB then MME shall need to forward paging request also towards SGSN in addition to sending it to eNb when this request is received via the SGs interface from the serving MSC/MSC server. This is required due the fact that MME is not able to determine whether the UE will be reachable via LTE or legacy 2G/3G radio access technologies.

5.8.4 Idle Mode versus Active Mode UE with CS Fallback for EPS

In similar fashion as today the LTE attached UE's EPS connection management (EMMECM) state in a packet switched domain is not known in detail by the serving MSC or MSC server providing circuit switched services for UE within circuit switched domain. This information is not provided to the circuit switched domain via the SGs interface and only MME or SGSN network elements have knowledge of this. Additionally due to this reason 3GPP has a specific non-EPS alert procedure which can be invoked by the MSC/MSC server (VLR) at any point in time to request via the SGs interface to EPS in order to be informed from events occurred in the EPS such as EMMECM state change. This procedure is especially used with the SMS use case (SMS over SGs) where it is used to eventually inform the short message service centre (SMSC) that the UE is again reachable via EPS.

Naturally after a call has been established via a circuit switched domain then the MSC or MSC server has knowledge of the state of the UE as of today.

The reason why this issue is important is related to the fact that 3GPP CSFB related specification defines a different kind of functional behaviour for EPS and UE depending on the state of the UE related to fallback performed using CSFB procedures. In case of SMS over SGs the UE state is not meaningful.

For the scope of this section the UE is considered to be in active mode when it has ongoing data transfer towards core network for reason or another. The UE is again considered to be in idle mode when it does not have ongoing data transfer or radio connectivity towards the core network.

In the case where the UE is in active mode then it has an ongoing radio resource connection with EPS and in this way is always considered to be reachable, thus paging is not required. In case fallback is performed, then it is possible that the UE is either redirected with RRC release procedure towards 2G/3G or PS HO procedure is performed as defined in (3GPP TS 23.401) and described in Section 4.2.3. Additionally in this case if the UE receives an incoming call from a circuit switched network then CSFB provides means for the network to identify the caller before fallback occurs by including a calling line identity parameter into the paging message that is sent towards the UE via the SGs interface by the MSC/MSC server. Based on this information the called subscriber is able to prevent fallback from occurring in the case where one does not want to receive the call.

In the case where the UE is in idle mode then in the case of an incoming, terminating circuit switched call the serving MME needs to perform paging via the S1 interface to reach the targeted UE subject for the call. Depending on the result of this paging procedure fallback can be performed or rejected towards the MSC or MSC server that performed paging.

The following example call scenarios describe CSFB from both a mobile originating call in idle mode to 2G and a mobile terminating call in active mode to 3G using PS HO procedure to move the ongoing data transfer to 3G.

5.8.5 CS Fallback Attachment

A prerequisite for the use of CSFB is that the UE has successfully performed combined EPS/IMSI attachment into the serving EPS and CS core network. This attachment is a normal EPS initial attachment (see Section 5.3) with CS attach specific additions as depicted in Figure 5.31.

1. The UE initiates an attachment as described in Section 5.3. The attach request message includes an indication for a combined EPS/IMSI attach and the MME allocates a location area identity (LAI) for the UE.
2. The MME derives a VLR number based on the allocated LAI and on an IMSI hash function.
3. The MME sends a location update request (LAI, IMSI, MME name, location update type) message to the MSC/VLR.

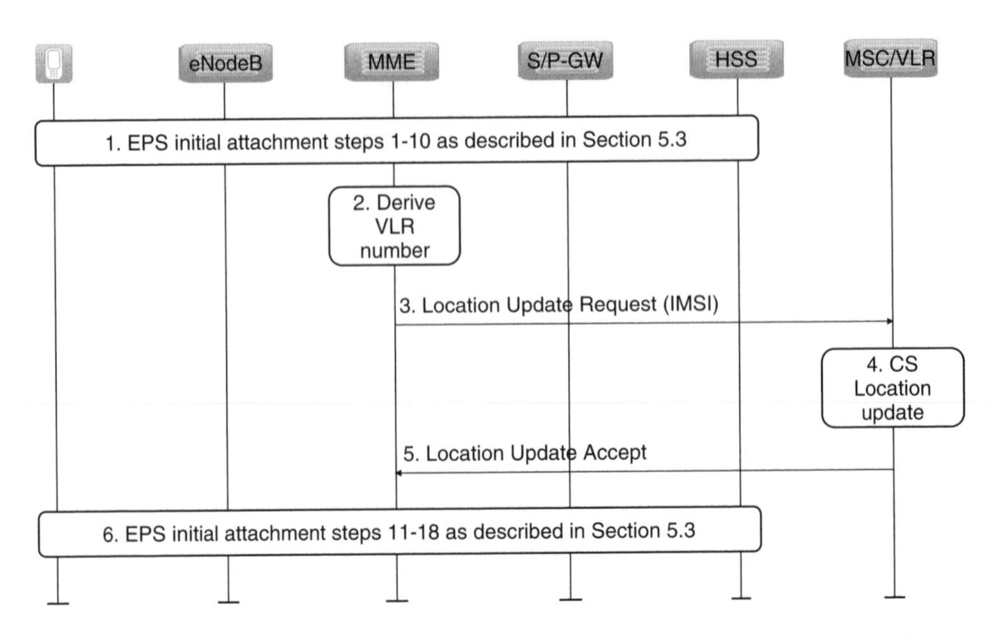

Figure 5.31 Combined EPS/IMSI attachment into serving EPS and CS core network.

4. The MSC/VLR performs a subscription check for CS and performs a location update procedure in the CS domain.
5. The MSC/VLR sends a location update accept to the MME.
6. The EPS attach procedure is completed by performing steps 11–18 as described in Section 5.3.

5.8.6 Mobile Originating Call Using CSFB

A mobile originating call (normal, emergency or priority) for an idle or active mode UE consists of a number of steps to move UE to 2G or 3G radio access and establish a voice call in an appropriate manner to the called destination.

Figure 5.32 represents the mobile originating call establishment in idle mode.

The CSFB procedure begins when triggered by the UE when it sends an extended service request via a NAS connection as defined in (3GPP TS 24.301) to the serving MME. This message contains an indication that CSFB is requested as described in Table 5.11.

An example of the relevant parts of a S1 extended service request is shown below (except id-TAI and id-EUTRAN-CGI parameters).

```
S1AP-PDU: initiatingMessage (0)
 initiatingMessage
  procedureCode: id-uplinkNASTransport (13)
  criticality: ignore (1)
  value
    UplinkNASTransport
      Item 1: id-NAS-PDU
        ProtocolIE-Field
```

```
        id: id-NAS-PDU (26)
        criticality: reject (0)
        value
          NAS-PDU: XYZ
          Non-Access-Stratum (NAS)PDU
            0010. . . .=Security header type: Integrity protected
                 and ciphered (2)
            . . . .0111=Protocol discriminator: EPS mobility
         management messages (7)
            Message authentication code: XYZ

            0000. . . .=Security header type: Plain NAS message, not
                 security  protected (0)
            . . . . 0111 = Protocol discriminator: EPS mobility
         management messages (7)
            NAS EPS Mobility Management Message Type: Extended
         service request (0x4c)
0. . . . . . . = Type of security context flag (TSC): Native security
      context (for KSIasme) (0)
            .000 . . . . = NAS key set identifier:  (0)
            . . . . 0000 = Service type: Mobile originating CS
         fallback or 1xCS fallback (0)
            Mobile identity - M-TMSI - TMSI/P-TMSI (0x1234567)
              Length: 5
              1111 . . . . = Unused
              . . . . 0. . . = Odd/even indication: Even number of
            identity digits
              . . . .100 = Mobile Identity Type: TMSI/P-TMSI (4)
              TMSI/P-TMSI: 0xc1234567
```

After this the MME will send a S1-AP INITIAL CONTEXT SETUP REQUEST procedure towards serving eNB to inform eNb that the terminal has requested invocation of the CSFB procedure. In the case where the UE would be in active mode instead then a modification would be triggered by EPC towards the UE. The relevant parts for CSFB of the S1-AP INITIAL CONTEXT SETUP REQUEST message is encoded as described in Table 5.12.

An example of the relevant parts of S1 initial context setup request is shown below.

```
S1AP-PDU: initiatingMessage (0)
  initiatingMessage
    procedureCode: id-initialContextSetup (9)
    criticality: reject (0)
    value
      Item 1: id-CSFallbackIndicator
        ProtocolIE-Field
          id: id-CSFallbackIndicator (108)
          criticality: reject (0)
          value
            CSFallbackIndicator: cs-fallback-required (0)
```

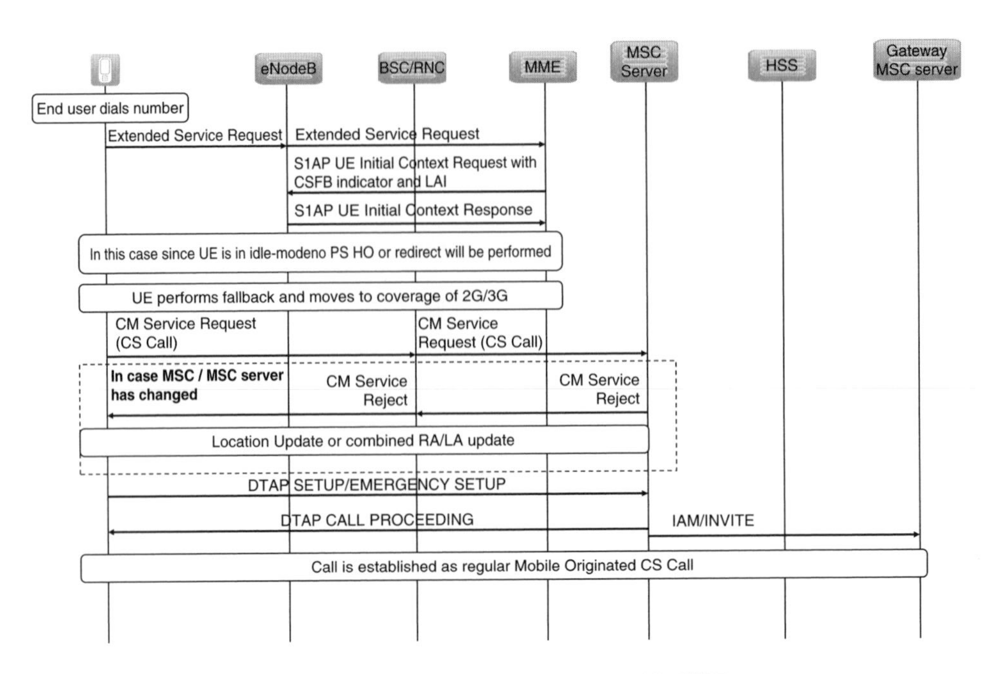

Figure 5.32 Mobile originating call with CSFB.

Table 5.11 Content of extended service request

Name of parameter	Purpose	Example value/content
Extended service request message identity	Identifies that the message is an extended service request	Value = 0×4C indicating an extended service request.
Service type	This parameter identifies that an extended service request was performed due to a mobile originating fallback for EPS by UE.	Value = 0×00 indicating a mobile originating call related CSFB request.

Table 5.12 Content of S1AP initial context setup request

Name of parameter	Purpose	Example value/content
CS fallback indicator	This parameter identifies that the served UE has requested invocation of either a normal CSFB or a priority call related CSFB.	'CSFB' or 'CSFB with high priority'

The eNb will reply back to the serving MME by using a S1-AP INITIAL CONTEXT SETUP RESPONSE message, which by itself can be understood to be positive acknowledgement for invocation of CSFB from the E-UTRAN point of view.

After the UE has been moved to the target 2G/3G radio access network it will start a normal mobile originating call establishment procedure by sending a CM service request message via GERAN/UTRAN to the MSC/MSC server. The MSC/MSC server serving the given radio access network may either accept the service request or reject it. In the case where it is rejected then the UE needs to perform a location update procedure and if accepted by the network then it continues mobile originating call establishment normally. In the latter case call establishment duration is increased a bit compared to the situation that no location update is performed, but this impact can be minimised by support for a follow on request (FoR) bit in location update sent by the UE towards the MSC/MSC server. This bit informs the circuit switched core network to keep the existing signalling connection towards radio access alive for the forthcoming call. Additionally UE may also include a circuit switched mobile originating (CSMO) flag into the same location update message to indicate that location update was related to an ongoing CSFB procedure. CSMO can be used by the core network for statistical purposes if supported.

It is also possible that the MSC/MSC server also rejects the location update procedure performed by the UE. In that case the UE shall consider call establishment as failed.

Table 5.13 Content of CM service request content of CM service request

Name of parameter	Purpose	Example value/content
CM service type	This parameter indicates the type of service requested by UE.	Value = 1 (mobile originated call or packet mode establishment)
Ciphering key sequence number	This parameter indicates the ciphering key proposed to be used by UE.	
Mobile station classmark 2	This parameter contains important information related to the UE's characteristics such as: Revision level (phase) of UE; Supported encryption algorithms (A5/1, A5/2 A5/3); Support for voice broadcast service, and so on.	As defined in (3GPP TS 24.008).
Mobile identity	This parameter contains the identity of the subscription using the UE. Also in the case of CSFB it is beneficial to use TMSI as much as possible, especially in the case of pooled (multipoint Iu/A)-based network architectures where TMSI contains a network resource identifier (NRI) value identifying the MSC or MSC server from pool.	TMSI or IMSI as defined in (3GPP TS 24.008).

In this example scenario the UE does not perform location update prior to performing a mobile originating call since the location area served by MSC/MSC server has not changed from the one that was used by the UE while connected to EPS.

Table 5.13 describes the content of the CM service request message relevant for call establishment purposes as defined by (3GPP TS 24.008).

An example of the relevant parts of the CM service request message is shown below when received from the A interface inside Direct Transfer Application Part (DTAP) message (i.e. fallback has been performed into GERAN).

```
GSM A-I/F DTAP - CM Service Request
    Protocol Discriminator: Mobility Management
         messages
    0000 .... = Skip Indicator: 0
    .... 0101 = Protocol discriminator: Mobility Management
         messages (5)
    00.. .... = Sequence number: 0
    ..10 0100 = DTAP Mobility Management Message Type: CM
         Service Request (0x24)

    Ciphering Key Sequence Number
    0... .... = Spare bit(s): 0
    .000 .... = Ciphering Key Sequence Number: 0

    CM Service Type
    .... 0001 = Service Type: (1) Mobile originating call
         establishment or packet mode connection establishment
    Mobile Station Classmark 2
    Length: 3
                 ...
```

After the connection management (CM) service request has been properly handled by the network then the UE will send a SETUP message to start call control procedures within the MSC/MSC server in order to route the call towards the called destination. The setup message is transferred transparently over the GERAN/UTRAN between the UE and MSC/MSC server.

Table 5.14 Setup message

Name of parameter	Purpose	Example value/content
Bearer capability, low-layer compatibility and high-layer compatibility	This parameter indicates the requested service from the network.	Speech, G3 facsimile, 3G-324M video telephony, and so on
Called party BCD number	This parameter identifies the called destination address.	E.164 address of called party
Supported codec list	This includes a list of the supported codecs when the UE supports a universal mobile telecommunications system (UMTS).	Wideband (WB) AMR and narrowband AMR

Table 5.14 represents a few selected parameters from the setup message which are important from a call establishment point of view.

An example of the relevant parts of the setup message is shown below when received from the A interface inside a DTAP message (i.e. fallback has been performed into GERAN).

```
GSM A-I/F DTAP - Setup
   Protocol Discriminator: Call Control; call related SS messages
   0... .... = TI flag: allocated by sender
   .000 .... = TIO: 0
   .... 0011 = Protocol discriminator: Call Control;
        call related SS messages (3)
   01.. .... = Sequence number: 1
   ..00 0101 = DTAP Call Control Message Type: Setup (0x05)
   Bearer Capability 1 <Contains information about nature of call;
        Speech, video
   etc.>

   Called Party BCD Number - (04012345678)
   Element ID: 94
   Length: 7
   1... .... .... ... = Extension: No Extension
   .000 .... = Type of number: unknown (0x00)
   .... 0001 = Numbering plan identification: ISDN/Telephony
        Numbering (Rec ITU-T E.164) (0x01)
   BCD Digits: 04012345678

   Supported Codec List
   Element ID: 64
   Length: 8
   System Identification (SysID): UMTS (0x04)
   Bitmap Length: 2
   Codec Bitmap for SysID 1
   <Supported codecs for UMTS>

   System Identification (SysID): GSM (0x00)
   Bitmap Length: 2
   Codec Bitmap for SysID 2
   <Supported codecs for GERAN>
```

All call related supplementary and intelligent network-based end user services and charging activity are performed by a circuit switched network normally as today. Also as visible in the example, call routing is performed by using existing NNIs (Network Network Interfaces), such as ISUP, BICC, SIP-I or native SIP as today [initial address message (IAM)/invite message in the figure above].

After the call is disconnected, then as in the case of a mobile originating call, the UE will perform idle-mode mobility based on instructions received from the serving radio access network and rules defined in (3GPP TS 23.221). Additionally it was later enhanced that MSC/MSC Server is able to inform serving BSC/RNC that subscriber's terminal has performed fallback due CSFB. It allows BSC/RNC to recognize those UEs that can be moved in more optimal way back to LTE. Also depending on capabilities of deployed system it may not be possible to transfer UE back to LTE sooner than all active PS connections are torn-down.

Via these mechanisms it is assumed that the UE is instructed to select LTE radio access as soon as possible, where coverage exists, in order to resume LTE end user experience.

5.8.7 Mobile Terminating Call Using CSFB

A mobile terminating call is a bit more complicated procedure than a mobile originating one from a network point of view. In this case, since the terminal is indirectly attached into the serving MSC/MSC server, that is there is an existing SGs association between the MME and VLR of the MSC/MSC server, the UE needs to be paged via the SGs interface towards EPS in order to trigger the fallback procedure.

In this example the mobile terminating circuit switched call is routed to a LTE attached UE in active mode, that is it has an ongoing packet switched connection via EPS. Therefore in addition to the fallback procedure a PS HO procedure will also be applied to move the terminal as swiftly as possible into target radio access to complete call establishment.

Figure 5.33 represents a mobile terminating call using CSFB in such a case when the MSC/MSC server is not changed during fallback.

Call establishment in the example above begins when the gateway MSC/MSC server receives an incoming call request (IAM or INVITE) from the originating network. This message contains information about the nature of the request [transport medium requirement (TMR) and user service information (USI)] as well as information about who (calling line identity) has called to whom (MSISDN of called subscriber).

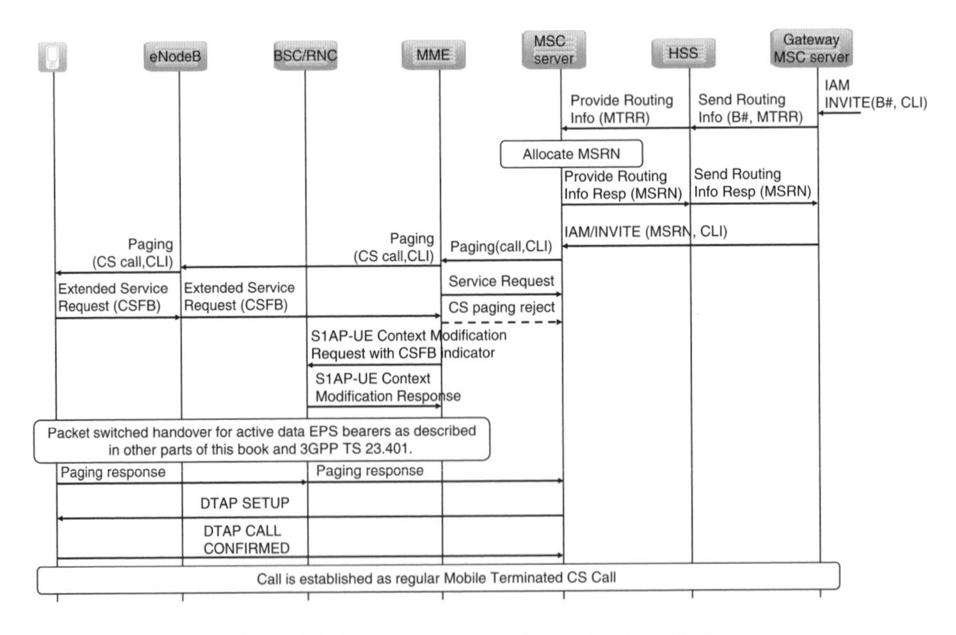

Figure 5.33 Mobile terminating call using CSFB.

The gateway MSC/MSC server performs a HLR (HSS) enquiry as described in (3GPP TS 23.018) in order to retrieve a mobile station roaming number (MSRN) from the visited MSC/MSC server having the SGs association for the called subscriber. This procedure uses send routing information message from the gateway MSC/MSC Server to HLR (HSS), which requests the MSRN from the VLR of the visited MSC/MSC server by using the provide roaming number (PRN) procedure. In this example the capability to use the MTRR procedure is shown in the send routing information and PRN messages in order to illustrate the way in which this capability is advertised between the gateway MSC/MSC server and the visited MSC/MSC server. Without this indication the visited MSC/MSC server shall not invoke the MTRR procedure.

Table 5.15 Content of SGsAP-paging-request

Name of parameter	Purpose	Example value/content
IMSI	This parameter contains the IMSI of the subscription to which this paging message is associated.	Identifier of subscription (15 digits long), including mobile country code, mobile network code and actual mobile subscriber identity number (MSIN)
TMSI	This parameter contains the TMSI allocated by the visited MSC/MSC server (VLR) for the subscriber.	TMSI (possibly including network resource identifier in case CS CN uses pooling)
Calling line identity	This parameter is sent by the visited MSC/MSC server when calling line identification (CLI) presentation has not been restricted by the caller. In the case where the UE is in active mode then MME will send the CLI as part of a paging request to the UE via an S1 interface. This information can be used by the called subscriber to reject possible fallback.	E.164 address of calling party
Service indicator	This parameter identifies the reason for the paging procedure. The reason can be either a CS call or a short message.	CS call or SMS
Location area identifier	This parameter includes information about the current location area served by the visited MSC/MSC server. In case fallback occurs to a different location area then the UE shall need to perform a location update procedure before it can receive an incoming call.	Location area as encoded in (3GPP TS 24.008)

Also not described in the figure both the gateway MSC/MSC server and HLR (HSS) can reside in a different network from the rest of the network elements. This is the case if the called subscriber has roamed to another network and has performed a combined IMSI/EPS attachment there.

After the call has been routed from the gateway MSC/MSC server to the visited MSC/MSC server then the visited terminating services are executed by the visited MSC/MSC server. These services include for instance blocking the identity of the calling subscriber in the case where the calling line identity restriction service has been requested by the caller. Additionally the visited MSC/MSC server may perform late (conditional) call forwarding services such as forward call when the called subscriber is busy in case such service needs to be invoked.

The paging procedure uses the SGsAP-PAGING-REQUEST message towards MME from the MSC/MSC server (VLR). Table 5.15 represents some key parameters included in the paging request.

An example of the relevant parts of SGsAP-paging-request message is shown below.

```
SGSAP Message Type: SGsAP-PAGING-REQUEST (0x01)
IMSI - IMSI (234123456789012)
  Element ID: 1
  Length: 8
  0010 . . . . = Identity Digit 1: 2
  . . . . 1. . . = Odd/even indication: Odd number of identity digits
  . . . .001 = Mobile Identity Type: IMSI (1)
  BCD Digits: 234123456789012
VLR name  -  VLR1.MSS1.MNC01.MCC234.3GPPNETWORK.ORG
  Element ID: 2
  Length: 38
  VLR name: VLR1.MSS1.MNC01.MCC234.3GPPNETWORK.ORG
Service indicator - CS call indicator
  Element ID: 32
  Length: 1
  Service indicator: CS call indicator (1)
TMSI - (0x200001)
  Element ID: 3
  Length: 4
  TMSI/P-TMSI: 0x00200001
Calling Party BCD Number - CLI - (358401234567)
  Element ID: 28
  Length: 8
  0. . . . . . . = Extension: Extended
  .001 . . . . = Type of number: International Number (0x01)
  . . . . 0001 = Numbering plan identification: ISDN/Telephony
       Numbering (Rec ITU-T E.164) (0x01)
  1. . . . . . . = Extension: No Extension
  .00. . . . = Presentation indicator: Presentation allowed (0x00)
  . . .0 00. . = Spare bit(s): 0
  . . . . .11 = Screening indicator: Network provided (0x03)
  BCD Digits: 358401234567
Location Area Identification (LAI)
  Element ID: 4
  Length: 5
```

```
  Location Area Identification (LAI) - 234/1/1000
  Mobile Country Code (MCC): United Kingdom of Great
       Britain and Northern Ireland (234)
  Mobile Network Code (MNC): Unknown (1)
  Location Area Code (LAC): 0x0064 (100)
Global CN-Id
  Element ID: 11
  Length: 5
  Mobile Country Code (MCC): United Kingdom of Great
       Britain and Northern Ireland (234)
  Mobile Network Code (MNC): Unknown (1)
  CN_ID: 100
eMLPP Priority
  Element ID: 6
  Length: 1
   0000 0. . . = Spare bit(s): 0x00
   . . . .111 = eMLPP Priority: Call priority level A (0x07)
```

After paging has been received by the UE then the MME will reply to the visited MSC/MSC server with a SGsAP-SERVICE-REQUEST message acknowledging the start of the paging procedure and to inform that a NAS connection has been successfully established with UE. Additionally the visited MSC/MSC server in this case is able to start a call forwarding no reply timer and provide an audible ringing tone indication towards the caller.

Table 5.16 describes the relevant parameters of SGsAP-SERVICE-REQUEST message.

An example of the relevant parts of SGsAP-SERVICE-REQUEST message is shown below.

```
SGSAP Message Type: SGsAP-SERVICE-REQUEST (0x06)
IMSI - IMSI (234123456789012)
  Element ID: 1
  Length: 8
  0010 . . . . = Identity Digit 1: 2
  . . . . 1. . . = Odd/even indication: Odd number of identity digits
  . . . .001 = Mobile Identity Type: IMSI (1)
  BCD Digits: 234123456789012
Service indicator - CS call indicator
  Element ID: 32
  Length: 1
  Service indicator: CS call indicator (1)
Tracking area identity
  Element ID: 35
  Length: 5
  Mobile Country Code (MCC): United Kingdom
       of Great Britain and Northern Ireland (234)
  Mobile Network Code (MNC): Unknown (1)
   Tracking area code(TAC): 0x0064
UE EMM mode
  Element ID: 37
  Length: 1
  UE EMM mode: EMM-CONNECTED (0)
```

Table 5.16 Content of SGsAP-service-request

Name of parameter	Purpose	Example value/content
IMSI	This parameter identifies the subscription to which this message is associated.	IMSI
Service indicator	This parameter identifies the reason for the paging procedure. The reason can be either a CS call or a short message.	CS call or SMS
IMEISV	This parameter contains IMEI and a software version of the UE that is being paged.	Identifier (16 digits long) containing an eight-digit type allocation code (TAC) identifying the terminal model, a six-digit serial number (SN) and a two-digit software version (SV)
UE time zone	This parameter contains the time zone of the UE as known by MME. This time zone can be different from the one used by the MSC/MSC server in the case where it handles time zones different from the MME.	As described in (3GPP TS 24.008)
Mobile station classmark 2	This parameter contains information related to characteristics of the UE.	As described in (3GPP TS 24.008)
Tracking area identifier	This parameter contains the last known tracking area identifier by MME.	As described in (3GPP TS 23.401)
E-UTRAN cell global identity	This parameter contains detailed information about the last known LTE cell of the UE known by MME.	As described in (3GPP TS 23.401).
UE EMM mode	This parameter indicates the current EPS mobility management (EMM) mode of the UE when the paging was initiated by the MSC/MSC server via an SGs interface.	EMM-connected EMM-idle

The UE will reply to the received paging request (for an incoming CS call) by using an extended service request message towards the serving MME. This message also contains an indication whether the called subscriber has accepted or rejected the incoming CS call already during the paging phase.

This enables the called subscriber to make a decision whether fallback occurs or not in the case where the called subscriber has an ongoing important data session via LTE and does not want that to be disturbed. In the case where the called subscriber has rejected the

Table 5.17 Content of S1 AP extended service request

Name of parameter	Purpose	Example value/content
Extended service request message identity	Identifies that message is extended service request	Value = 0×4C indicates an extended service request.
Service type	This parameter identifies that an extended service request was performed due to amobile originating fallback for EPS by UE.	Value = 0 × 01 indicates a mobile terminating call related CSFB request.

incoming call then MME will indicate this towards the visited MSC/MSC server by using a SGsAP-PAGING-REJECT message including the SGs cause value 'mobile terminating CS fallback call rejected by the user'.

Table 5.17 describes important parameters in the extended service request message as described in (3GPP TS 24.301) when the called subscriber has accepted an incoming call attempt.

An example of the relevant parts of the S1 AP extended service request message is shown below.

```
S1AP-PDU: initiatingMessage (0)
  initiatingMessage
    procedureCode: id-uplinkNASTransport (13)
    criticality: ignore (1)
    value
      UplinkNASTransport
       . . .
       Item 1: id-NAS-PDU
         ProtocolIE-Field
         id: id-NAS-PDU (26)
         criticality: reject (0)
         value
           Non-Access-Stratum (NAS)PDU
             0010 . . . . = Security header type: Integrity
                  protected and ciphered (2)
             . . . . 0111 = Protocol discriminator: EPS mobility
                  management messages (7)
             . . .
             0000 . . . . = Security header type:Plain NAS
                  message,not security protected (0)
             . . . . 0111 = Protocol discriminator: EPS mobility
                  management messages (7)
             NAS EPS Mobility Management Message Type: Extended
                  service request (0x4c)
             . . .
             . . . . 0001 = Service type: Mobile terminating CS
                fallback or 1xCS fallback (1)
             CSFB response
```

```
                1011 .... = Element ID
                .... 0... = Spare bit(s): 0x00
                .....01 = CSFB response: CS fallback accepted
                    by the UE (1)
```

Since the UE is in active mode while performing fallback it means that the MME will also need to configure the ongoing EPS connections towards eNb. In this case this procedure is started by sending a S1AP UE context modification request message towards eNB from the MME. The message is encoded as described in (3GPP TS 36.413). This message contains also CS fallback indicator indicating to eNb that fallback will be soon performed. eNodeb will reply to the MME with a S1AP UE context modification response as acknowledgement for the request.

After this exchange of signalling messages has been completed between eNb and MME the EPS begins to move the active EPS bearer connections to the target radio access network (2G/3G). In this example it is assumed that PS HO is used towards 3G radio access technology. Execution of the PS HO procedure is described in Section 5.6.1. In addition to standard PS HO procedure eNb will inform the target RNC by using the source RNC to the target RNC transparent container that also CSFB has occurred. Also in the case where a priority call is in question then that can be also indicated in the same way to target the RNC.

When the UE has performed the fallback to target radio access (in this example UTRAN) then it will reply by sending a paging response via RNC to the visited MSC/MSC server. In the case where the location area would be changed during the fallback procedure then the UE shall need to perform a location update to the visited MSC/MSC server, which also could have been changed. In this example it is assumed that the MSC/MSC server and location area remains the same.

The visited MSC/MSC server will send a setup message towards the UE after receiving the paging response. This setup contains sufficient information for the UE to have information about the nature of the call. Also calling line identity is provided in order to inform the called subscriber about the caller's identity. Table 5.18 presents some key parameters in the setup message as described in (3GPP TS 24.008).

An example of the relevant parts of SETUP message is shown below.

```
GSM A-I/F DTAP - Setup

   Protocol Discriminator: Call Control; call related SS messages
0... ... = TI flag: allocated by sender
.000 .... = TIO: 0
.... 0011 = Protocol discriminator: Call Control;
     call related SS messages (3)
00.. .... = Sequence number: 0
..00 0101 = DTAP Call Control Message Type: Setup (0x05)
Bearer Capability 1  <Contains information about nature of
     call: Speech, Video, etc.>
...
Calling Party BCD Number - (358040123456)
  Element ID: 92
  Length: 8
```

```
0... ... = Extension: Extended
.001 .... = Type of number: International Number (0x01)
.... 0001 = Numbering plan identification: ISDN/Telephony
    Numbering (Rec ITU-T E.164) (0x01)
1... ... = Extension: No Extension
.00. ... = Presentation indicator: Presentation allowed (0x00)
...0 00.. = Spare bit(s): 0
.....11 = Screening indicator: Network provided (0x03)
BCD Digits: 358040123456
```

The UE will reply to setup with a CALL CONFIRMED message by also indicating a list of supported speech codecs, which then is used by the visited MSC/MSC server to select the appropriate speech codec for the call.

Finally the call is completely established between the caller and the called subscriber. After the call is disconnected, then as in the case of a mobile originating call, the UE will perform idle-mode mobility based on instructions received from the serving radio access network and rules defined in (3GPP TS 23.221). Additionally it was later enhanced that MSC/MSC Server is able to inform serving BSC/RNC that subscriber's terminal has performed fallback due CSFB. It allows BSC/RNC to recognize those UEs that can be moved in more optimal way back to LTE. Also depending on capabilities of deployed system it may not be possible to transfer UE back to LTE sooner than all active PS connections are torn-down.

Via these mechanisms it is assumed that the UE is instructed to select LTE radio access as soon as possible, where coverage exists, in order to resume LTE end user experience.

5.8.8 Call Unrelated CSFB Procedures

In addition to basic voice and video telephony use cases which utilise fallback from LTE to 2G/3G the 3GPP has also standardised the following non-call related procedures which may be required in practical CSFB deployments.

- Mobile initiated call-independent supplementary service procedure;
- Network initiated call-independent supplementary service procedure;
- Mobile originating location request (MO-LR) procedure;
- Mobile terminating location request (MT-LR) procedure.

Table 5.18 Content of setup message

Name of parameter	Purpose	Example value/content
Bearer capability, low-layer compatibility and high-layer compatibility	This parameter indicates a service requested from the network.	Speech, G3 facsimile, 3G-324M video telephony, and so on
Calling party BCD number	This parameter identifies the calling line identity of the caller.	E.164 address of calling party

These procedures are all based on the same mode of operation in which the UE will perform fallback from LTE to 2G/3G after which the actual procedure will then be performed via the existing circuit switched core network. After the procedure has been completed, then the UE can re-attach into LTE by using idle-mode mobility principles defined in (3GPP TS 23.221) and if sufficient LTE coverage exists.

Both mobile and network initiated call-independent supplementary service procedures are applicable when the UE wants to control supplementary service settings via a traditional UE user interface or when the network has supplementary service-specific information which need to be transmitted to UE. In the former situation this procedure can be used for instance to activate a supplementary service, interrogate the state of supplementary service or deactivate a supplementary service. These procedures are defined in (3GPP TS 24.010). It should be noted that when a terminal is capable for Ut/XCAP interface-based service configuration then the UE shall need to use Ut/XCAP as the primary mechanism for that purpose.

In the case of a MO-LR procedure based on UE capabilities either fallback will be performed or alternatively EPC procedures can be used instead to retrieve the current location of the UE. The UE is able to determine whether EPC supports the EPC-MO-LR procedure by checking the content of the location control services (LCS) support indicator provided during combined IMSI/EPS attachment phase from EPC. In the case where the LCS support indicator only indicates that the network supports the CS-MO-LR procedure then the UE shall perform fallback to 2G/3G access to retrieve the location. In the case where the EPC-MO-LR procedure is supported or the UE has an ongoing IMS-based voice telephony session then the UE should avoid fallback procedure.

The MT-LR procedure is triggered by a gateway mobile location centre (GMLC) that has received a request to retrieve the current location of a subscriber from the serving mobile network. In the case where the UE has performed a combined IMSI/EPS attachment and has registered to a circuit switched domain via EPS then HLR (HSS) has the VLR address of subscriber pointing to the MSC/MSC server having a SGs association for that particular subscriber. This SGs association, as defined in an earlier chapter of this book, is used to perform paging by sending a SGsAP-paging-request with an LCS indicator including that paging is due to a MT-LR procedure. Please see more details related to the content of the LCS identifier from (3GPP TS 29.118).

In all mobile terminating procedures previously described in this section, if the UE performs fallback from LTE to 2G/3G and if the visited MSC/MSC server happens to be different than one that performed original paging, then the procedure will fail. This is due to the reason that 3GPP has not standardised any procedure similar to the ones described in the next chapter which would re-route the mobile terminating procedure towards a new MSC/MSC server.

In today's mobile networks external applications may also use MAP Any Time Interrogation (ATI) procedure to interrogate current location of user. This ATI procedure is used towards HLR (HSS), which again uses MAP Provide Subscription Information (PSI) towards serving VLR respectively. Initially support for PSI in case user has active SGs association was not defined by 3GPP. There has been recent proposals to enhance this situation which are not described in more detail in this book. However issue need to be taken into account when planning the deployment of CSFB in case such applications exists in network.

5.8.9 *Mobile Terminating Roaming Retry and Forwarding*

In the case when the UE performs fallback due mobile terminating call from LTE to 2G/3G but the target radio access network is controlled by a MSC/MSC server different from the one that originally performed the paging via the SGs interface towards EPS, then the UE shall perform a location update to the new MSC/MSC server, as described in Section 5.8.7.

This location update may contain a CSMT flag that indicates to the new MSC/MSC server controlling the 2G/3G radio access network that a location update procedure has been invoked during the ongoing fallback procedure. The CSMT flag in the new MSC/MSC server enables it to keep an existing signalling connection towards the UE open for forthcoming call establishment from the gateway MSC/MSC server which is expected to arrive shortly. In the case where CSMT is not supported either by the UE or a new MSC/MSC server then a new paging procedure need to be invoked and thus the call establishment time will be longer than with a CSMT flag.

After location update has been performed properly then the HLR (HSS) of the called subscriber will cancel the subscription from the old MSC/MSC server by using a MAP CANCEL LOCATION message. In this case, depending on what kind of procedure is supported by the circuit switched network (both serving and home PLMN), then the following events will happen.

5.8.9.1 Mobile Terminating Roaming Retry

In the case where the MTRR procedure is supported end to end as defined in (3GPP TS 23.018) and (3GPP TS 23.272) then the old MSC/MSC server will send a MAP RESUME CALL HANDLING (RCH) message towards the gateway MSC/MSC server. This message indicates that a MTRR procedure has been requested by the old MSC/MSC server from the gateway MSC/MSC server. A prerequisite for the use of MTRR is that the gateway MSC/MSC Server has indicated the support for that procedure via HLR (HSS) to the old MSC/MSC server as described earlier (see Section 5.8.7). Sending of a MAP RCH message is triggered by the reception of a MAP cancel location from HLR (HSS).

In the case where the gateway MSC/MSC server decides to invoke a MTRR procedure then it re-executes a new HLR enquiry towards HLR (HSS) in order to retrieve a MSRN, as described in (3GPP TS 23.018) and (3GPP TS 23.272). This is the second HLR enquiry that will be performed for the given call by the same gateway MSC/MSC server network element. The gateway MSC/MSC server also releases ongoing IN dialogues started from the gateway basic call state model (BCSM) and re-establishes those based on a new HLR enquiry.

As the result of this procedure the call is re-routed from the gateway MSC/MSC server to a new MSC/MSC server that controls the target 2G/3G radio access network to which the UE has performed fallback.

Figure 5.34 describes at high level the execution of a MTRR procedure.

Use of this procedure requires that both the home PLMN as well as the serving PLMN, if different, support the MTRR and a sufficient MAP version to transfer RCH message from the old MSC/MSC server to the gateway MSC/MSC server. This may be a new requirement for an inter-PLMN MAP interface. No support is required for MTRR procedure from the target (new) MSC/MSC server that controls the 2G/3G radio access to which the fallback was performed.

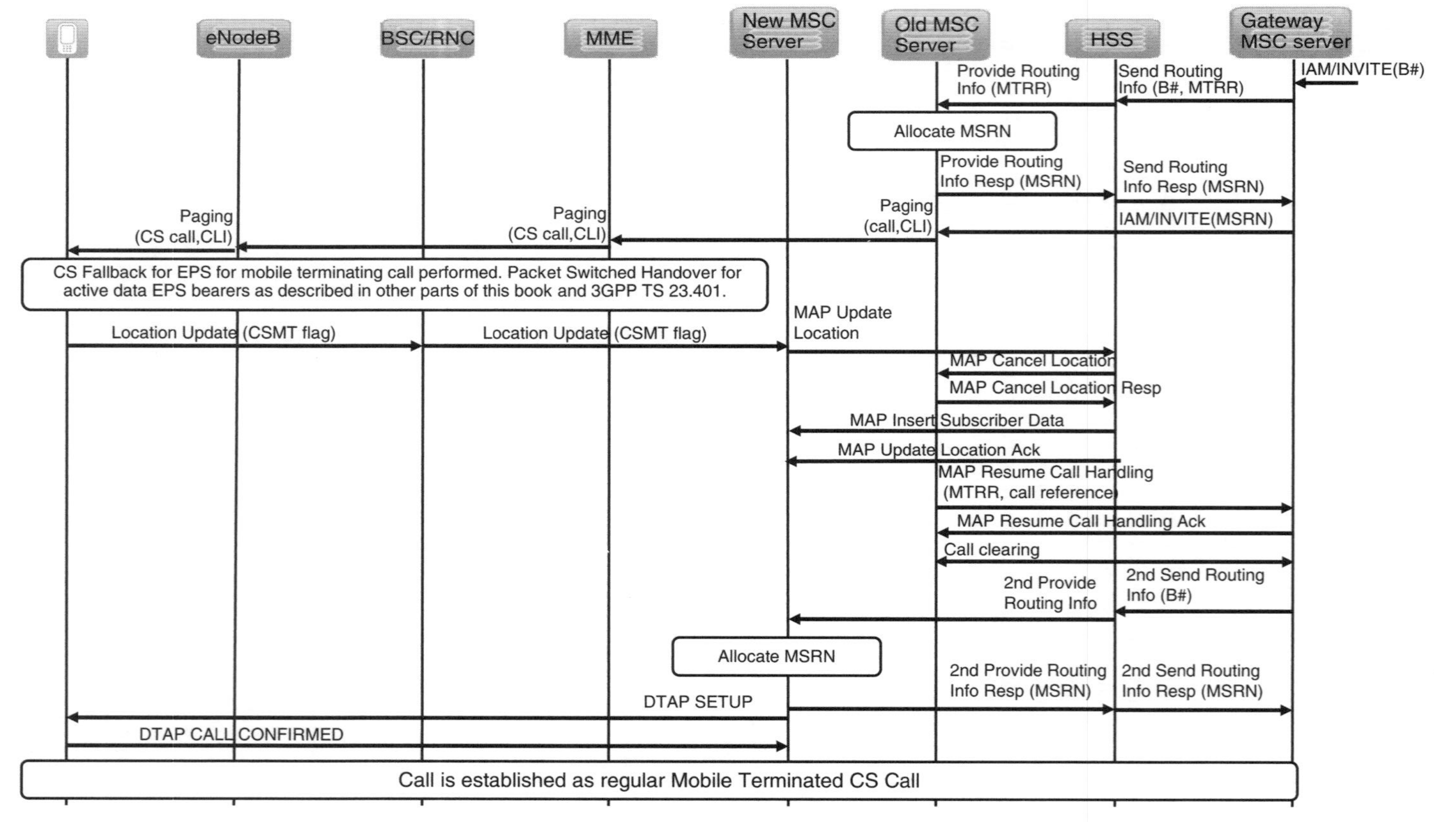

Figure 5.34 Mobile terminating call using MTRR procedure.

Table 5.19 Content of MAP resume call handling

Name of parameter	Purpose	Example value/content
Call reference	Identifies the call in question.	Maximum string of eight octets long byte.
All information sent	This parameter indicates that all parameters are sent from the visited MSC (VMSC) to the gateway MSC.	Presence of parameter fulfils the condition.
MT roaming retry	This parameter indicates that the reason for sending MAP resume call handling message was due invocation of MT roaming retry procedure.	Presence of parameter fulfils the condition.

Table 5.19 describes the content of a MAP RCH message.

Due the reason that the MTRR procedure requires wide support from circuit switched networks, 3GPP has standardised another newer mechanism called mobile terminating roaming forwarding as part of 3GPP Release 10. This procedure is more optimised in the sense that it only requires support for it from the serving PLMN and it has no impact for the gateway MSC/MSC server. Additionally the call is no longer re-routed from the home PLMN to the serving PLMN, causing a lower call setup time since the call can be re-routed entirely in the serving PLMN as described below.

5.8.9.2 Mobile Terminating Roaming Forwarding

The triggering event for the MTRF procedure is the same as with MTRR as described above and described in (3GPP TS 23.018).

After receiving the location update procedure from the UE that has performed the fallback procedure and in the case where the UE has indicated a CSMT flag in this procedure as described above then the new MSC/MSC server that controls 2G/3G radio access network indicates MTRF capability as part of a MAP SEND IDENTIFICATION message towards the old MSC/MSC server that performed the paging via SGs towards EPS.

After this the new MSC/MSC server also performs location update normally to HLR (HSS) which again cancels the subscription from the old MSC/MSC server as described above in case of MTRR. The old MSC/MSC server, also supporting a MTRF procedure, will send a MAP PRN message towards the new MSC/MSC server as a new functionality that is unique for the MTRF procedure. In this way the new MSC/MSC server will allocate a MSRN, as in the case of a normal HLR enquiry performed by the gateway MSC/MSC server, and will provide it to the old MSC/MSC server in response. The old MSC/MSC server will route the call directly to the new MSC/MSC server by using this number within a serving PLMN, in this way optimising the call setup time as well as removing the need to support re-routing via a gateway MSC/MSC server as would be case with the MTRR procedure.

Figure 5.35 describes at high level the execution of a MTRF procedure.

The use of MTRF requires support from both the old and new MSC/MSC servers located in the serving PLMN. This is a difference, compared to the MTRR in which both the old MSC/MSC server having a SGs interface and the gateway MSC/MSC

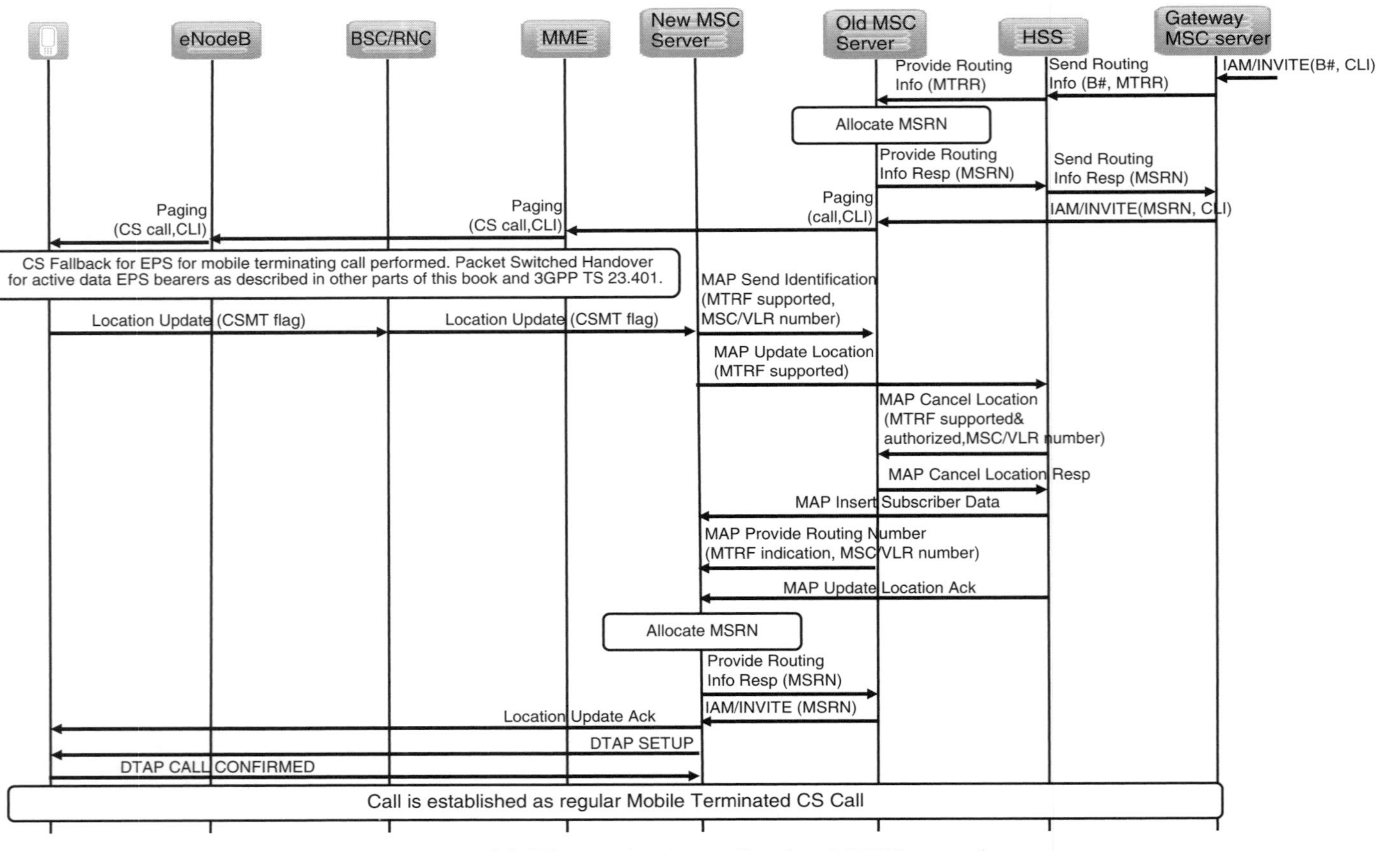

Figure 5.35 Mobile terminating call using MTRF procedure.

server network elements involved in call establishment process require support for the MTRR procedure.

Depending on whether the UE uses IMSI or TMSI while performing the location update procedure to a new MSC/MSC server then HLR (HSS) support is either required or not. In the case where the UE uses TMSI then HLR (HSS) support is not required for MTRF and a new MSC/MSC server can resolve the address of the old MSC/MSC server based on TMSI. However in the case where the UE uses IMSI then a new MSC/MSC server need to indicate a requirement for MTRF to the MAP UPDATE LOCATION message that will be sent to HLR (HSS). HLR (HSS) needs to send this parameter to the old MSC/MSC server in order to invoke the sending of a MAP PRN to the new MSC/MSC server, as described in the Figure 5.35. Thus in the most optimal case UEs would always use TMSI as part of their location update procedure to a new MSC/MSC server and then no HLR (HSS) support would be required.

In summary the mechanism to prevent the use of MTRF from a home PLMN is standardised and therefore is likely to be implemented.

5.8.9.3 Co-Existence of MTRR and MTRF Procedures

It is also possible, due to the fact that MTRR and MTRF procedures can, in a phased manner, be independently introduced into different networks, that both procedures are simultaneously available for use for the same call in both the home and serving PLMN.

Selection of the procedure call to be used will be based on the local O&M configuration and capabilities existing in the network elements in the serving PLMN.

5.8.10 Summary

This section describes CSFB functionality for EPS that enables the operator to re-use existing circuit switched core network infrastructure to provide voice and SMS on top of LTE data services.

It is expected that initial deployments for CSFB will be made for SMS cases but eventually also voice, video telephony as well as other use cases will be deployed into the network. CSFB enables operators also to gradually migrate to IMS-based VoIP in parallel providing support for capabilities for use cases such as emergency call, roaming and priority call that may not be natively enabled at that time with IMS-based VoIP. Therefore it is highly expected that CSFB will be used in parallel to IMS-based VoIP, even with the same UEs. Also it may in some cases be the selected approach that SMS is still deployed by using CSFB procedures (SMS over SGs/LTE) even when the voice and other services would be based on IP. Naturally in the long run it is expected that IMS-based VoIP will be the only single technology in use but the transition period for that may be long.

In addition to the basic call flow descriptions of example CSFB scenarios, this chapter also describes different alternative technologies that can be used on the EPS side to perform fallback as well as to move active packet switched connections to the target radio access network (2G/3G).

Finally radio access related details such as features that can be used to improve end to end performance can be found in Section 4.2.3.

5.9 VoLTE Messaging

Messaging services have been available for many years in mobile networks and have contributed a significant deal of operator revenues coming directly from the services available in today's mobile networks. According to some public studies, for instance in Finland, the number of short messages sent within the six month time period from January 2010 to June 2010 was as huge as over 1900 million short messages sent (this means 700 SMS annually per each inhabitant in Finland) whereas in same time period the number of multimedia messages was as high as 17 million messages (source: FICORA). The reason to use Finland as an example was selected since Finland has the one of the highest mobile penetrations in the world as well as one of the fastest reductions in the number of plain old telephony fixed access lines.

Operator provided messaging services are considered essential services in VoLTE for both consumer and businesses, even though complementary technologies have been introduced since the introduction of SMS. These include instant messaging introduced as part of the rich communication suite and many Internet-based services such as Skype, Microsoft Messenger or even more highly integrated solutions within UE messaging applications such as Google Talk, Apple iMessage and What's App services, which all are assumed to have significant impact to end user messaging behaviour. These services will probably gradually take over users from the more traditional short and multimedia messaging services (MMSs) but that will likely consume a lot of time, and the speed of this will vary based on markets.

Additionally one issue that reduces this movement towards pure-IP based messaging services is related to the services that end users are able to use by using SMS today. It is possible for instance to order a bus ticket or to buy a bottle of lemonade by using a short message and actual payment will be based on the end user's mobile phone bill. This is a strong capability that cannot be easily replaced by Internet service providers since instant messaging and other IP-based messaging applications have more often been considered as free of charge services rather than trusted, payable services having a strong local connection to services close to subscribers.

In addition to basic short message and multimedia messages USSD has also been included into this analysis, which is also being used widely today for a number of different service use cases such as topping up the pre-paid accounts, requesting dial-in from a home network when roaming with a pre-paid subscription as well as controlling services in a home network without a native phone user interface, to name a few examples.

5.9.1 Native IMS Messaging

Two different types of native IMS messaging forms exist: immediate messaging and session-based messaging. Each form of IMS messaging has its own characteristics; so, even though messaging in its simplest form can be thought of as a single service – after all, all forms of messaging are really about sending a message from A to B – the fact that these characteristics differ makes them each a service on their own. However, the way in which applications are built on top of these services may well hide the fact that these are different forms of messaging.

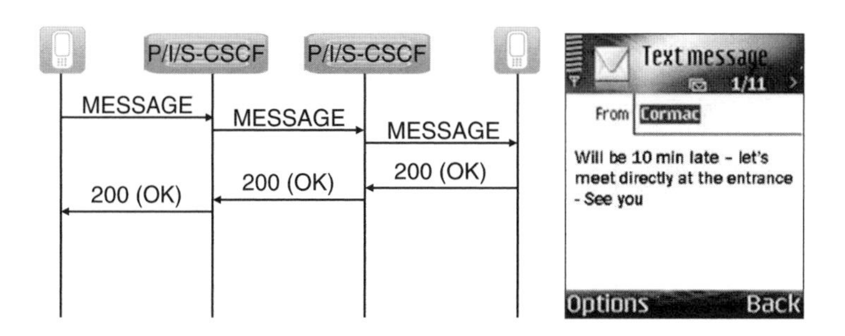

Figure 5.36 Immediate messaging flow.

5.9.1.1 Immediate Messaging

Immediate messaging, or page-mode messaging, is the familiar instant messaging paradigm adopted in the IMS framework. It uses the SIP MESSAGE method to send messages between peers in near-real time. Figure 5.36 illustrates a typical message flow. In immediate messaging, the UE simply generates a MESSAGE request, fills in the desired content – which typically consists of text, but can also contain snippets of multimedia such as sounds and images – and populates the Request-URI with the address of the recipient. The request is then routed via the IMS infrastructure similar to the manner used for an INVITE (see Section 5.5.2), until the immediate message finds its way to the UE of the recipient user or gets stored in the network. There might, of course, be a reply to this message; in fact, a full dialogue of immediate messages back and forth between the two users is likely. However, in contrast to session-based messaging, the context of this session only exists in the minds of the participating users. There is no protocol session involved: each immediate message is an independent transaction and is not related to any previous requests.

5.9.1.2 Session-Based Messaging

In this mode of messaging the user takes part in a session in which the main media component often consists of short textual messages. As in any other session a message session has a well-defined lifetime: a message session starts when the participants begin the session and stops when the participants close the session. After the session is set up – using SIP and SDP between the participants – media then flows directly from peer to peer. Figure 5.37 illustrates the typical message flow of a message session. Session-based messaging can be peer to peer, in which case the experience closely mimics that of a normal voice call. An ordinary invitation to a session is received, the only difference being that the main media component is a session of messages. However, this is not an actual limitation to session-based messaging, since it is, of course, possible to combine other media sessions with message sessions. In fact, many useful and exciting applications are enabled by this functionality: for example video calls with a text side channel might be a valuable application for hearing-impaired people. The actual protocol for conveying the messages within a session is called the message session relay protocol (MSRP) (RFC4975). MSRP layers on top of TCP, and can carry any multipurpose

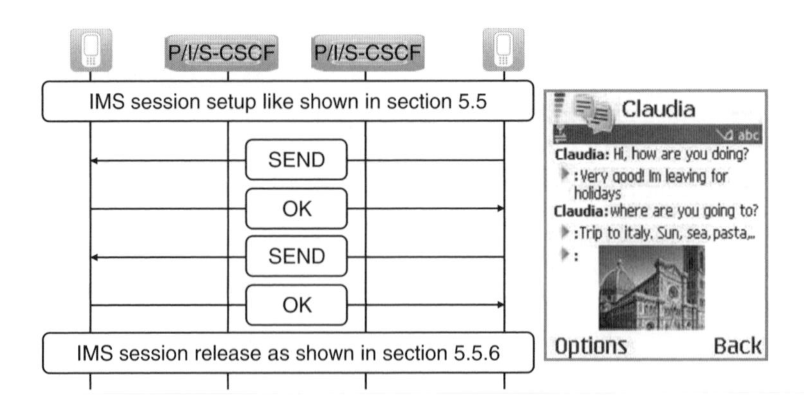

Figure 5.37 Session-based messaging flow.

Internet mail extensions (MIMEs) encapsulated data. Messages can be of arbitrary size, since one of the protocol features is the ability to support sending a complete message in small chunks that are automatically reassembled at the recipient end. Session-based messaging forms a natural unison with conferencing as well. Using the conferencing functionality, session-based messaging can turn into a multiparty chat conference.

In this mode of operation, session-based messaging can enable applications similar to modern-day voice conferences. A service provider will typically offer the possibility for users to have both private chats, where the set of participants is restricted and public chats, some of which are maintained by the service provider. In this context the network typically provides an additional functionality called MSRP switch. It is used to relay messages between participants.

5.9.2 SMS Interworking

SMS has been a tremendous success in the past and still continues to be such in the LTE era. Lots of different services have been developed in such a fashion all across the globe that use SMS as the enabling technology. Therefore continuation of this kind of such service is extremely important for the LTE era as well. To deliver the actual SMS messages to the UE in LTE access 3GPP has standardised two solutions: SMS over SGs and SMS over IP. In addition 3GPP has defined AS functionality to translate SMS to IMS immediate messaging and vice versa.

In both ways to provide SMS it is important to note that the common data model of relay layer protocol (RP) is used to transfer the actual short message payload even though the transport is completely different (SIP protocol vs SGsAP/S1AP protocol). This is shown in Figure 5.38.

The CP-data message is transferred by using a short message control protocol (SM-CP) and is used in the CM layer to exchange an upper layer protocol such as short message relay protocol (SM-RP) and transfer protocol data unit (TPDU). RP-data messages are transferred by using SM-RP used in the short message relay layer (SM-RL). The actual

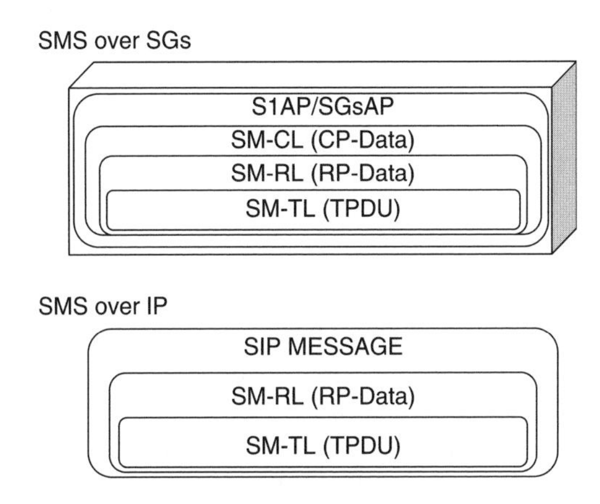

Figure 5.38 Protocol layers of SMS over SGs/IP.

application content of the short message is transmitted as a TPDU that is part of the short message transfer layer (SM-TL).

More details of the protocol layer architecture for short message can be found from (3GPP TS 23.040) and (3GPP TS 24.011) specifications.

For example, in the case of SMS over SGs the CM layer is part of the transferred protocol content in the SGs interface. In such a case the CP-data, CP-ack and CP-error messages are embedded inside the NAS message container (3GPP TS 29.118).

The CP-data message embedded inside the NAS message container contains the following parameters, as described in Table 5.20.

CP-data is acknowledged by the MSC server serving the given subscriber with the CP-ack message. Alternatively CP-error can be used to inform the UE of any error which possibly occurred during the procedure.

Table 5.20 Content of CP-data

Name of parameter	Purpose	Example value/content
Protocol discriminator	This parameter contains a description of the protocol payload included in the CP-data.	Value 'SMS messages' indicates the presence of SMS payload.
Transaction identifier	Identifies the transaction in question between the two endpoints involved in messaging.	Value encoded as in (3GPP TS 24.007).
Message type	Actual purpose of this message.	Value can be either CP-DATA, CP-ACK or CP-ERROR.
CP-user-data	Contains RP-data container.	See Table 5.21 for content of RP-data.

An example of a CP-data message is as follows.

```
GSM A-I/F DTAP - CP-DATA
  Protocol Discriminator: SMS messages
  DTAP Short Message Service Message Type: CP-DATA (0x01)
  CP-User Data
  Extraneous Data
```

CP-user data contains the relay protocol data that contains an encapsulated RP-data message as defined below. RP-data is the first common message definition that is also used with SMS over IP as encapsulated inside the SIP MESSAGE payload.

In addition to RP-data, also RP-ack, RP-error and RP-SMMA messages have been specified in (3GPP TS 24.011). The RP-ack message is used for acknowledgement of successfully received RP-data whereas RP-error can be used to inform a peer entity that RP-data has not been successfully processed. The RP-SMMA message is used by the UE to inform SMSC that the UE has again memory available to receive new short messages.

Table 5.21 contains a description of information encoded inside RP-data.

An example of an RP-data message is as shown below.

```
GSM A-I/F RP - RP-DATA (MS to Network)
  Message Type RP-DATA (MS to Network)
  RP-Message Reference
    RP-Message Reference: 0x01 (1)
  RP-Origination Address
    Length: 0
  RP-Destination Address - (358401234567)
     Length: 7
     1... .... = Extension: No Extension
     .001 .... = Type of number: International Number (0x01)
     .... 0001 = Numbering plan identification: ISDN/Telephony
          Numbering (Rec ITU-T E.164) (0x01)
     BCD Digits: 358401234567
  RP-User Data
     Length: 16
     TPDU (not displayed)
```

The TPDU container is transferred transparently between UE and SMSC. Depending on the type of TPDU it can include one of the following protocol messages, according to Table 5.22.

The scope of this section covers important messages such as SMS-DELIVER and SMS-SUBMIT with the respective acknowledgement message, which are represented in the following tables.

In the case where the UE originates the short message, then SMS-SUBMIT is encoded by the UE as shown in Table 5.23.

In the case where the UE receives the short message then the originating SMSC will send the SMS-DELIVER message towards the UE, encoded as shown in Table 5.24.

Table 5.21 Content of RP-data

Name of parameter	Purpose	Example value/content
RP-message type	This parameter contains a type of message.	Message type value can be either RP-data, RP-ack, or RP-error
RP-message reference	This parameter contain a sequence number to associate the content of the RP-data with a certain message transaction.	Range between 0 and 255
RP-originator address	In the case of a mobile terminating transaction this identifies the address of the sending SMSC.	E.164 address
RP-destination address	In the case of a mobile originating transaction this identifies the address of the receiving SMSC.	E.164 address
RP-user data	Contains a transport layer protocol data unit (TPDU) container	See Table 5.22 for content of TPDU

Table 5.22 Content of TPDU

Name of message	Purpose
SMS-deliver	Used to transfer a short message from SMSC to UE.
SMS-submit	Used to transfer a short message from UE to SMSC.
SMS-deliver-report	Used to transfer the indication of an error or positive acknowledgement as a response to for instance an SMS-deliver message.
SMS-submit-report	Used to transfer an indication of an error or positive acknowledgement as a response to for instance an SMS-submit message.
SMS-status-report	Status report from SMSC to UE.
SMS-command	Command from UE to SMSC.

As a response to the SMS-DELIVER message the SMS-DELIVER-REPORT message is encapsulated within the CP-ack message towards the SMSC.

An example of a TPDU container encoded as part of SMS-SUBMIT is shown below:

```
GSM SMS TPDU (GSM 03.40) SMS-SUBMIT
  0... .... = TP-RP: TP Reply Path parameter is not set in this SMS
SUBMIT/DELIVER
  .0.. ..= TP-UDHI: The TP UD field contains only the short message
  ..0.. ... = TP-SRR: A status report is not requested
  ...0 0... = TP-VPF: TP-VP field not present (0)
  ... ...0.. = TP-RD: Instruct SC to accept duplicates
  ... ...01 = TP-MTI: SMS-SUBMIT (1)
  TP-MR: 18
  TP-Destination-Address - (358509876543)
    Length: 12 address digits
```

```
    1... ... :  No extension
    .001 .... :  Type of number: (1) International
    .... 0001 :  Numbering plan: (1) ISDN/telephone (E.164/E.163)
    TP-DA Digits: 358509876543
  TP-PID: 0
    00... ...:  defines formatting for subsequent bits
    ..0.. ...:  no telematic interworking, but SME-to-SME protocol
    ...0 0000 :  the SM-AL protocol being used between the SME
    and the MS (0)
  TP-DCS: 0
    00... ... = Coding Group Bits: General Data Coding
         indication (0)
    Special case, GSM 7 bit default alphabet
  TP-User-Data-Length: (3) depends on Data-Coding-Scheme
  TP-User-Data
    SMS text: Hello World!
```

5.9.2.1 Short Message Service over SGs

SMS will be made available for different kinds of LTE-capable UEs. The first kind of UEs will be such that are in fact USB or integrated data modems having no legacy CS capability at all. Therefore full-scale CS fallback for EPS functionality is not required by such UEs but instead only SMS functionality. From the end user perspective there is some vendor or communication service provider branded application that can be executed within a PC and which can be used to either send or more often receive short messages. Additionally it is possible to send short messages for over the air configuration purposes to name some use cases that do not involve the end user. This means that the communication

Table 5.23 Content of SMS-submit

Name of parameter	Purpose	Example value/content
SMS flags	Parameter contains flags set by sender to modify treatment related to short message.	Features that can be requested by UE:Status report;Same reply path for message. Same reply path for message.
TP-destination-address	This parameter contains the MSISDN address of the recipient of a short message.	MSISDN address.
TP-protocol-identifier	This parameter contains an identifier of the used protocol.	In the case of a textual short message this is encoded into 0 value.
TP-data coding scheme	This parameter contains information about encoding of the payload.	In the case of a textual short message this is encoded into 0 value (seven bit encoding).
TP-user data	This parameter contains a textual representation of the short message payload.	Actual text.

Table 5.24 Content of SMS-deliver

Name of parameter	Purpose	Example value/content
SMS flags	Parameter contains flags set by sender to modify treatment related to short message.	Features that can be requested by UE: Status report; Same reply path for message.
TP-originating-address	This parameter contains the MSISDN address of the sender of a short message.	MSISDN address.
TP-protocol-identifier	This parameter contains the identifier of the used protocol.	In case of textual short message this is encoded into 0 value.
TP-data coding scheme	This parameter contains information about encoding of the payload.	In the case of a textual short message this is encoded into 0 value (seven bit encoding).
TP-service-centre-time-stamp	This parameter contains a timestamp set by the SMSC, which indicates the time the message was sent towards the UE.	Value contains year, month, day, minutes and seconds as well as timezone information.
TP-user data	This parameter contains a textual representation of the short message payload.	Actual text.

service provider is able to remotely configure the behaviour of the USB data modem in a similar fashion as in case of UTRAN/GERAN enabled USB data modems today.

In the case where the UE also has CS capabilities for voice and other CS services then it is likely the short message user interface will be exactly the same as that used by the end user when the UE is attached to UTRAN/GERAN radio access. Thus the LTE can be considered as yet another transport for old SMS.

Additionally it is likely that, in the first LTE deployments when LTE roaming capability is also introduced, then communication service providers will face same kind of requirements (possible regulatory ones) to provide SMS in order to inform roaming customers about cost of roaming data service and also inform when the cost of service exceeds a certain threshold. This is known as 'roaming bill shock prevention' and is currently mandated by the European Union (EU).

In order to support the delivery of short message over LTE as signalling information in a similar fashion as a short message is delivered today in existing mobile networks, changes have been introduced to all relevant signalling interfaces between the UE and the serving MSC or MSC server network element.

In the LTE-Uu interface between UE and eNb, the short message is transferred transparently within a NAS protocol defined in (3GPP TS 24.301). This interface is between the UE and MME. The MME then forwards the content to serving MSC or MSC server that supports the SGs interface as defined in (3GPP TS 29.118). In the (3GPP TS 24.011) specification this architectural model is described as a circuit switched service in S1 mode indicating that the UE is connected to a circuit switched SMS via the S1 interface. The short message content can be either a single or a concatenated payload and both mobile originated and terminating short message are supported.

Support for SMS over SGs does not require any specific functionality from the eNb element since the actual payload is transferred as NAS signalling transparently between the LTE attached UE and the serving MME.

Current understanding in the industry is that a sufficient standardisation baseline for SMS delivery within LTE signalling (NAS) is 3GPP Release 8 and no improvements are required from later 3GPP releases.

Following chapters will describe the way in which a short message payload can be transferred via a circuit switched domain by using SMS over SGs (TS 23.272) functionality.

A pre-requirement for these procedures is that the LTE UE has performed successful combined IMSI/EPS attachment as described in Section 5.8.5.

Mobile Originating Short Message: Mobile originating short message delivery is initiated by the UE by performing an UE triggered service request towards an EPC. After this event the UE is able to send an uplink NAS transport message to MME which gets mapped into an uplink unitdata message of the SGs interface towards the serving MSC or MSC server. These uplink messages contain the actual payload of the short message.

The serving circuit switched network element, that is the MSC or MSC server, will perform the needed tasks for the mobile originated short message in a similar fashion to case of a message that would have been received from the UE via GERAN/UTRAN radio access. This means that, for example the required intelligent network services as well as charging can be performed at this point.

Eventually the message is forwarded to the SMSC of the served subscriber via a MAP interface, which then completes the short message delivery in a normal fashion towards the originating UE.

Figure 5.39 describes the message flow of an originating SMS delivery via LTE radio access using SMS over SGs.

Uplink and downlink signalling messages in each key interface contain the actual payload data of a short message that is a NAS message container.

The following example represents one of the key messages SGsAP-uplink-unitdata which is used to exchange actual short message payload within the SGs interface between the EPS and the circuit switched domain. Content of SGsAP-downlink-unitdata is considered as similar and thus not shown here separately.

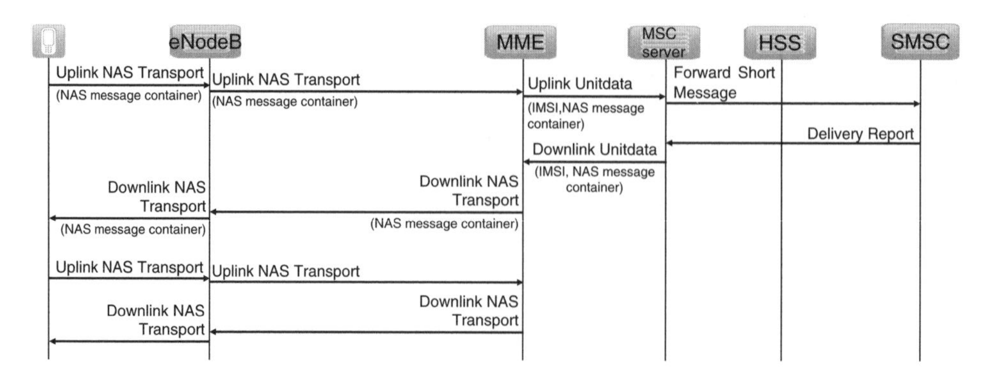

Figure 5.39 Mobile originating short message over SGs.

Other interfaces are left for the reader to investigate, since MME performs only parameter level conversion between the SGs and S1 interface without touching the actual content.

The actual message content of a SGsAP-uplink-unitdata message, which is used to transfer the short message payload from UE to circuit switched domain, is described in Table 5.25.

An example of a SGsAP-uplink-unitdata message is shown below.

```
SGSAP Message Type: SGsAP-UPLINK-UNITDATA (0x08)
IMSI - IMSI (234051234567890)
    Element ID: 1
    Length: 8
    0010 .... = Identity Digit 1: 2
    .... 1... = Odd/even indication: Odd number of identity
       digits
    .....001 = Mobile Identity Type: IMSI (1)
    BCD Digits: 234051234567890
NAS message container
    Element ID: 22
    Length: 2
    GSM A-I/F DTAP - CP-ACK
        Protocol Discriminator: SMS messages
            1... .... = TI flag: allocated by receiver
            .000 .... = TIO: 0
            .... 1001 = Protocol discriminator: SMS messages (9)
        DTAP Short Message Service Message Type: CP-ACK (0x04)
        Message Elements
Tracking area identity
    Element ID: 35
    Length: 5
    Mobile Country Code (MCC): United Kingdom of Great
 Britain and Northern Ireland   (234)
    Mobile Network Code (MNC): Unknown (100)
    Tracking area code(TAC): 0x0064
```

Table 5.26 describes again the content of a SGsAP-downlink-unitdata message, which is used to transfer a short message payload from a circuit switched domain towards the UE.

The next section will describe the mobile terminating short message.

Mobile Terminating Short Message: Mobile terminating short message delivery is a slightly more complicated procedure than the mobile originating one.

In this case, since the UE is indirectly attached into the serving MSC or MSC server (CS core network) by MME, then the SMSC of the sender will route the short message by using a MAP interface to that particular MSC or MSC server having SGs association for the served recipient. This procedure is handled in exactly same fashion as a short message delivery in mobile networks today.

A MSC or MSC server having SGs association will perform the required services execution tasks for a mobile terminating short message in a similar fashion as is performed today if the subscriber is registered via GERAN/UTRAN radio access. After these tasks

Table 5.25 Content of SGsAP-uplink-unitdata

Name of parameter	Purpose	Example value/content
IMSI	This parameter contains the IMSI of the served subscriber.	Identifier of subscription (15 digits long), including mobile country code, mobile network code and actual mobile subscriber identity number (MSIN).
NAS message container	This parameter contains the actual short message service payload including the actual message.	Byte string encoded as defined in (3GPP TS 24.011).
IMEISV	This parameter contains the identity of the subscriber's terminal (IMEI) as well as a software version.	Identifier (16 digits long) containing eight-digit type allocation code (TAC) identifying the terminal model, six-digit serial number (SN), and two-digit software version (SV).
UE time zone	This parameter contains the current timezone of the subscriber's UE. It should be noted that this value can be different than the timezone of the serving circuit switched domain in the case where the serving LTE access network is located in a different timezone.	Value indicates the difference between GMT and local time.
Mobile station classmark 2	This parameter contains classmark information including:Support for MT SMS;Support for A5/1 algorithm.	This information is structure encoded as defined in (3GPP TS 24.008).
TAI	This parameter contains the current tracking area identifier set of the UE, if available in MME.	Value contains mobile country code, mobile network code and tracking area code of serving TAI
E-CGI	This parameter contains the current evolved cell global identity of the UE, if available in MME.	Value contains mobile country code, mobile network code and cell identifier of serving LTE cell.

Table 5.26 Content of SGsAP-downlink-unitdata

Name of parameter	Purpose	Example value/content
IMSI	This parameter contains the IMSI of the served subscriber.	Identifier of subscription (15 digits long), including mobile country code, mobile network code and actual mobile subscriber identity number (MSIN).
NAS message container	This parameter contains the actual short message service payload, including the actual message.	Byte string encoded as defined in (3GPP TS 24.011).

the MSC or MSC server will perform paging via the SGs interface. In the case of successful paging the actual short message will be delivered via the SGs interface within the DL unitdata message to the MME, which again will forward the payload transparently to the UE with DL NAS transport message. Finally delivery of the short message will be completed by exchanging the delivery report from the UE to the SMSC of the sender and acknowledging completion of the procedure towards the UE of the recipient.

Figure 5.40 represents the message flow of a terminating short message delivery via LTE radio access using SMS over SGs.

As described above, a mobile terminating short message delivery begins by paging the UE via LTE access. This paging is requested by a circuit switched network by sending a SGsAP-paging-request message encoded as disclosed in Table 5.27.

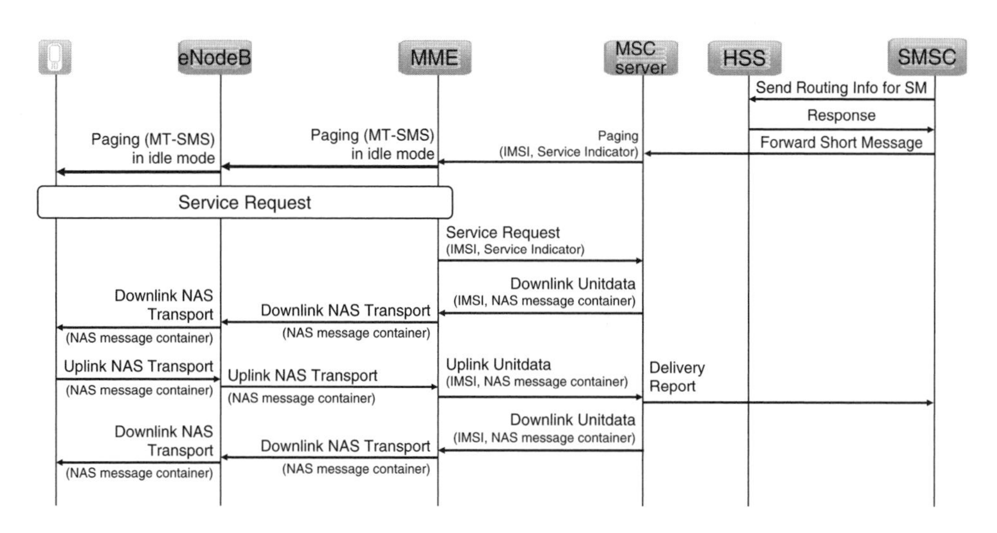

Figure 5.40 Mobile terminating short message over SGs.

Table 5.27 Content of SGsAP-paging-request

Name of parameter	Purpose	Example value/content
IMSI	This parameter contains the IMSI of the served subscriber.	Identifier of subscription (15 digits long), including mobile country code, mobile network code and actual mobile subscriber identity number (MSIN).
Service indicator	This parameter indicates the type of service that is being requested from EPS	
	Possible values include:CS call indicator;SMS Indicator.	SMS indicator = 0×02

An example of a SGsAP-paging-request message is shown below.

```
SGSAP Message Type: SGsAP-PAGING-REQUEST (0x01)
IMSI - IMSI (234051234567890)
    Element ID: 1
    Length: 8
    0010 . . . . = Identity Digit 1: 2
    . . . . 1. . . = Odd/even indication: Odd number of identity
       digits
    . . . . .001 = Mobile Identity Type: IMSI (1)
    BCD Digits: 234051234567890
Service indicator - SMS indicator
    Element ID: 32
    Length: 1
    Service indicator: SMS indicator (2)
VLR name  -  VLR01.MSS02.MNC001.MCC234.3GPPNETWORK.ORG
    Element ID: 2
    Length: 41
    VLR name: VLR01.MSS02.MNC001.MCC234.3GPPNETWORK.ORG
Global CN-Id
    Element ID: 11
    Length: 5
    Mobile Country Code (MCC): United Kingdom of
     Great Britain and Northern Ireland (234)
    Mobile Network Code (MNC): Unknown (100)
    CN_ID: 100
Location Area Identification (LAI)
    Element ID: 4
    Length: 5
    Location Area Identification (LAI) - 234/100/1000
        Mobile Country Code (MCC): United Kingdom of Great
             Britain and Northern Ireland (234)
        Mobile Network Code (MNC): Unknown (100)
```

```
        Location Area Code (LAC): 0x03e8 (1000)
    TMSI - (0x20000)
        Element ID: 3
        Length: 4
        TMSI/P-TMSI: 0x00020000
```

After receiving the SGsAP-PAGING-REQUEST from the serving circuit switched domain the MME will acknowledge this by sending a SGsAP-SERVICE-REQUEST message with content as defined in Table 5.28.

An example of a SGsAP-service-request is shown below.

```
    SGSAP Message Type: SGsAP-SERVICE-REQUEST (0x06)
    IMSI - IMSI (234051234567890)
          Element ID: 1
          Length: 8
          0010 .... = Identity Digit 1: 2
          .... 1... = Odd/even indication: Odd number of
          identity digits
          .....001 = Mobile Identity Type: IMSI (1)
  BCD Digits: 234051234567890
    Service indicator - SMS indicator
          Element ID: 32
          Length: 1
          Service indicator: SMS indicator (2)
    Tracking area identity
          Element ID: 35
          Length: 5
          Mobile Country Code (MCC): United Kingdom
              of Great Britain and Northern Ireland (234)
          Mobile Network Code (MNC): Unknown (100)
          Tracking area code(TAC): 0x0064
```

The content of both the SGsAP-uplink-unitdata and SGsAP-downlink-unitdata messages are the same as shown in the previous chapter. As in the case of the previous

Table 5.28 Content of SGsAP-service-request

Name of parameter	Purpose	Example value/content
IMSI	This parameter contains the IMSI of the served subscriber.	Identifier of subscription (15 digits long), including mobile country code, mobile network code and actual mobile subscriber identity number (MSIN).
Service indicator	This parameter indicates the type of service that is being requested from EPS Possible values include:CS call indicatorSMS indicator.	SMS indicator = 0×02

section, other interfaces are left for the reader to investigate since MME performs only parameter level conversion between the SGs and S1 interface without touching the actual content.

5.9.2.2 SMS over IP

The SMS over IP solution is primarily targeted for UEs that use IMS-based VoIP and it also provides SMS capability to UEs that are attached to non-cellular IP CANs [e.g. wireless local area network (WLAN) and WiMAX]. The SMS over IP solution is very straightforward; the actual SMS is attached as a special content type to the SIP MESSAGE method. This enables the UE to send and receive SMS over the IMS network. Utilising this type of interworking allows all kinds of existing value added SMS services to be delivered to users connected to the IMS. Figure 5.41 shows an example of SMS over IP origination flow.

At step 1 UE composes and sends a SIP MESSAGE. Key fields at the MESSAGE request are: Request-URI that points to SMS Service Centre (recipient's address is included in the actual short message) and Content-Type stating that the payload is an actual short message. The MESSAGE gets routed to IP-SM-GW due to initial filter criteria (see also Section 4.5.5). At step 8 the IP-SM-GW extracts the actual short message and sends it forward. Steps 9–14 depict an optional short message delivery report procedure.

Figure 5.42 shows a terminating SMS over IP procedure. At step 1 IP-SM-GW receives SMS from the SMS Service Centre and it composes and sends a SIP MESSAGE in step 2. Key fields at the MESSAGE request are: (i) Request-URI that points to a registered IMS public user identity, (ii) Accept-Contact stating that this request can only be delivered to a UE that has registered capability to receive SMS over IP, (iii) Request-Disposition having

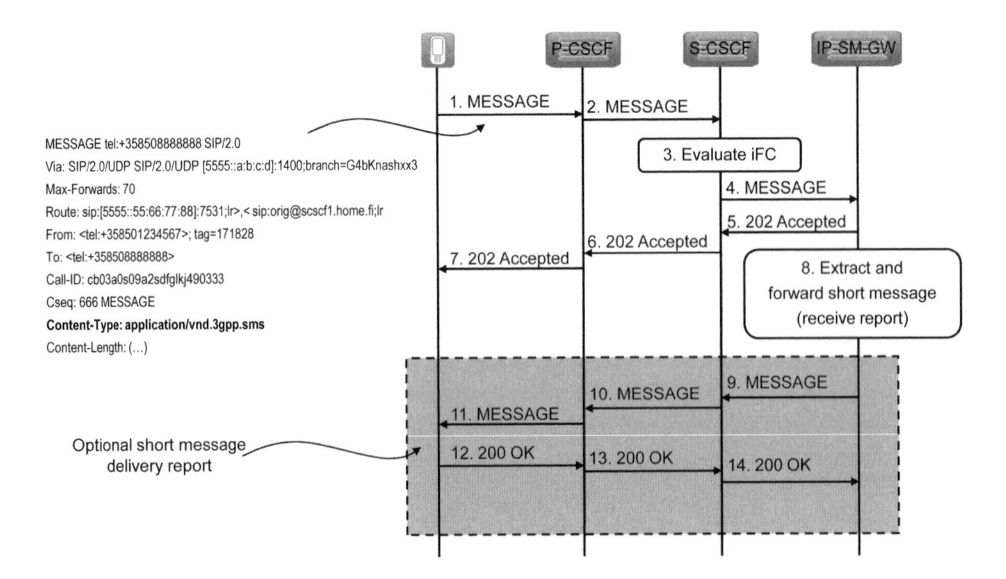

Figure 5.41 Example of originating SMS over IP.

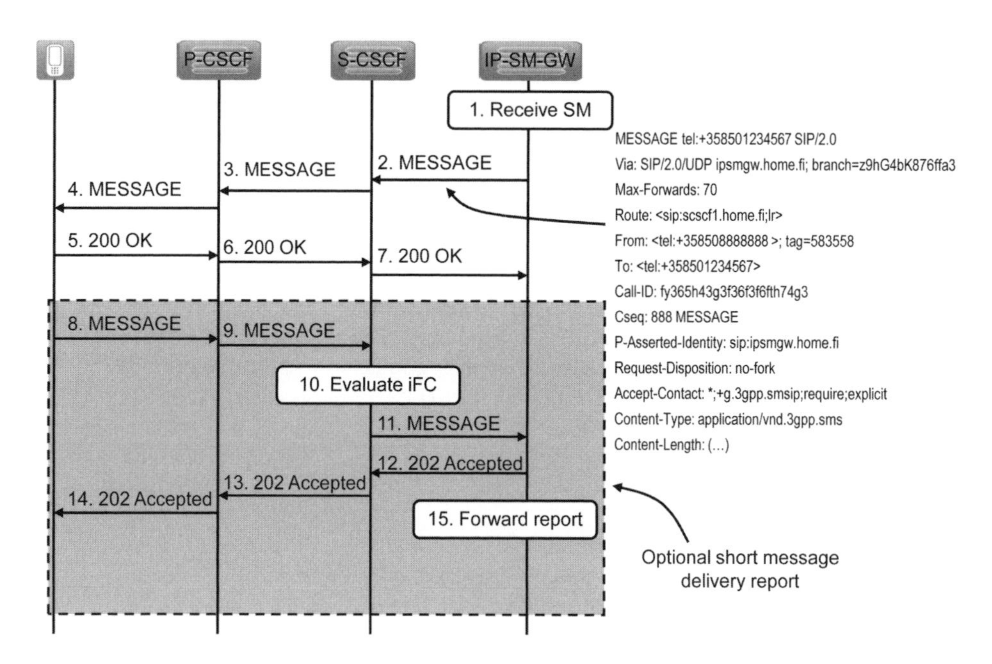

Figure 5.42 Example of terminating SMS over IP.

the value 'no-fork' stating that this request is to be delivered to one and only one UE and (iv) Content-Type stating that the message payload is an actual short message (3GPP TS 24.341). Steps 8–15 depict an optional short message delivery report procedure.

Let us analyse the protocol differences of SMS over IP and legacy text-based SIP messaging described in Section 5.9.1.1. with a concrete example. At first Kristiina's VoLTE compliant UE (+358501234567) sends a SMS over IP '*Will be 10 minutes late – let's meet directly at the entrance – See you*' to her friend (+358405556667) having also VoLTE compliant UE and IMS subscription. The SIP MESSAGE from Kristiina's UE will look like this:

```
MESSAGE tel:+358508888888 SIP/2.0
Via: SIP/2.0/UDP [5555::a:b:c:d]:1400; branch=abc123
Max-Forwards: 70
Route: <sip:[5555::55:66:77:88]:7531;lr>,<sip:orig@scscf1.home.fi;lr>
P-Access-Network-Info:3GPP-E-UTRAN-TDD;utran-cell-id-
3gpp=2440051F3F5F7
From: <tel:+358501234567>; tag=583558
To: <tel:+358508888888>
Call-ID: fy365h43g3f36f3f6fth74g3
Cseq: 888 MESSAGE
Require: sec-agree
Proxy-Require: sec-agree
Security-Verify: ipsec-3gpp; alg=hmac-sha-1-96; spi-c=98765432;
  spi-s=87654321; port-c=8642; port-s=7531
Content-Type: application/vnd.3gpp.sms
```

```
Content-Length: ...
GSM A-I/F RP - RP-DATA (MS to Network)
  Message Type RP-DATA (MS to Network)
  RP-Message Reference
     RP-Message Reference: 0x03 (3)
  RP-Origination Address
     Length: 0
  RP-Destination Address - (358508888888)
     Length: 7
     1... .... = Extension: No Extension
     .001 .... = Type of number: International Number (0x01)
     .... 0001 = Numbering plan identification: ISDN/Telephony
       Numbering (Rec ITU-T E.164) (0x01)
     BCD Digits: 358508888888
  RP-User Data
     Length: 160
     TPDU (not displayed)
GSM SMS TPDU (GSM 03.40) SMS-SUBMIT
     0... .... = TP-RP: TP Reply Path parameter is not set
       in this SMS SUBMIT/DELIVER
     .1.. .... = TP-UDHI: The beginning of the TP UD field
       contains a Header in addition to the short message
     ..0. .... = TP-SRR: A status report is not requested
     ...0 1... = TP-VPF: TP-VP field present - enhanced format (1)
     .... .0.. = TP-RD: Instruct SC to accept duplicates
     .... ..01 = TP-MTI: SMS-SUBMIT (1)
  TP-MR: 3
  TP-Destination-Address - (358405556667)
     Length: 12 address digits
     1... .... :  No extension
     .001 .... :  Type of number: (1) International
     .... 0001 :  Numbering plan: (1) ISDN/telephone (E.164/E.163)
     TP-DA Digits: 358405556667
  TP-PID: 0
     00.. .... :  defines formatting for subsequent bits
     ..0. .... :  no telematic interworking, but SME-to-SME protocol
     ...0 0000 :  the SM-AL protocol being used
       between the SME and the MS (0)
  TP-DCS: 0
     00.. .... = Coding Group Bits: General Data Coding
       indication (0)
     Special case, GSM 7 bit default alphabet
  TP-Validity-Period
     0... .... :  No extension
     .0.. .... :  Not single shot SM
     ..00 0... :  Reserved
     .... .010 :  Validity Period Format: relative
     255 seconds
  TP-User-Data-Length: (160) depends on Data-Coding-Scheme
  TP-User-Data
     User-Data Header
```

```
        User Data Header Length (5)
        IE: Concatenated short messages, 8-bit reference
          number (SMS Control)
           Information Element Identifier: 0
           Length: 3
           Message identifier: 2
           Message parts: 2
           Message part number: 1
           10.. .... :  Fill bits
         SMS-text: Will be 10min late - let's meet directly
           at the entrance - See you
```

Now the recipient (also having a VoLTE compliant UE with IMS subscription) responds to this message 'Ok. I'll wait for you at the entrance'.

```
MESSAGE tel:+358401112223 SIP/2.0
Via: SIP/2.0/UDP [5555::d:c:b:a]:1357;branch=z9hG4bKna
Max-Forwards: 70
Route: <sip:[6666::33:44:55:66]:6701;lr>,
  <sip:orig@scscf1.home1.net;lr>
P-Access-Network-Info:3GPP-E-UTRAN-TDD;utran-cell-id-3gpp=2449145353
From: <tel:+358405556667>; tag=171828
To: <tel:+358401112223>
Call-ID: cb03a0s09a2sdfglkj55555
Cseq: 666 MESSAGE
Require: sec-agree
Proxy-Require: sec-agree
Security-Verify: ipsec-3gpp; alg=hmac-sha-1-96;
spi-c=2345456; spi-s=345346346; port-c=2342; port-s=6701
Content-Type: application/vnd.3gpp.sms
Content-Length: ...
 GSM A-I/F RP - RP-DATA (MS to Network)
   Message Type RP-DATA (MS to Network)
   RP-Message Reference
    RP-Message Reference: 0x03 (3)
   RP-Origination Address
    Length: 0
   RP-Destination Address - (358401112223)
    Length: 7
    1. . . . . . .. = Extension: No Extension
    .001 .... = Type of number: International Number (0x01)
    .... 0001 = Numbering plan identification: ISDN/Telephony
      Numbering (Rec ITU-T E.164) (0x01)
    BCD Digits: 358401112223
   RP-User Data
    Length: 160
    TPDU (not displayed)
 GSM SMS TPDU (GSM 03.40) SMS-SUBMIT
   0... ..... = TP-RP: TP Reply Path parameter
     is not set in this SMS SUBMIT/DELIVER
```

```
  .1... .... .. = TP-UDHI: The beginning of the TP UD field
    contains a Header in addition to the short message
  ..0... .... .. = TP-SRR: A status report is not requested
  ...0 1... = TP-VPF: TP-VP field present - enhanced format (1)
  .... .0.. = TP-RD: Instruct SC to accept duplicates
  .... ..01 = TP-MTI: SMS-SUBMIT (1)
  TP-MR: 3
  TP-Destination-Address - (358501234567)
   Length: 12 address digits
   1... .... .: No extension
   .001 .... : Type of number: (1) International
   .... 0001 : Numbering plan: (1) ISDN/telephone (E.164/E.163)
   TP-DA Digits: 358501234567
  TP-PID: 0
   00... .... : defines formatting for subsequent bits
   ..0... .. : no telematic interworking, but SME-to-SME protocol
   ...0 0000 : the SM-AL protocol being used between the SME
     and the MS (0)
  TP-DCS: 0
   00... .... = Coding Group Bits: General DataCoding indication (0)
   Special case, GSM 7 bit default alphabet
  TP-Validity-Period
   0... .... : No extension
   .0.. .... : Not single shot SM
   ..00 0... : Reserved
   .... .010  : Validity Period Format: relative
   255 seconds
  TP-User-Data-Length: (160) depends on Data-Coding-Scheme
  TP-User-Data
   User-Data Header
     User Data Header Length (5)
     IE: Concatenated short messages, 8-bit reference
       number (SMS Control)
        Information Element Identifier: 0
        Length: 3
        Message identifier: 2
        Message parts: 2
        Message part number: 1
     10... ....: Fill bits
SMS-text: Ok. I'll wait for you at the entrance
```

This same example is repeated in the following section but in that case the sender of the short message uses a circuit switched phone and conversion to IMS subscriber is performed.

5.9.2.3 SMS – SIP Messaging Interworking

In addition to the SMS over SGs and SMS over IP 3GPP has specified interworking between SMS and a native SIP-based messaging (see Section 5.9.1) solutions. It means that SMS is fully converted to a SIP-based request, meaning that IMS UE no longer needs to implement a SMS stack. This work can be seen as an evolutionary step from the SMS over IP feature. As a continuation of the previous example, a user with circuit switched phone (+358503333333) sends the same SMS 'Will be 10 minutes late – let's meet directly at the entrance – See you' to her friend (+358405556667). As the recipient is an IMS user the terminating SMS gets routed to IP-SM-GW serving the recipient which converts the short message to a SIP MESSAGE, for example as follows[15]:

```
MESSAGE tel:+ 358405556667 SIP/2.0
Via: SIP/2.0/UDP ipsmgw.home1.net; branch=z9hGSttg4egs
Max-Forwards: 70
Route: <sip: scscf1.home1.net;lr>
From: <tel:+358503333333>; tag=54365
To: <tel:+358405556667>
Call-ID: ghreh56346363
Cseq: 785 MESSAGE
P-Asserted-Identity: tel:+358503333333
Content-Type: plain/text
Content-Length: 67
Will be 10min late - let's meet directly at the entrance -  See you
```

The key differences compared to the Message request carrying SMS over IP (Figure 5.42) are: the actual message is included in text format, the request shows that the request is coming from A-party and Accept-Contact and Request-Disposition headers are not set. When the recipient gets this request she will not be able to see that the original request was SMS. Let us assume that the IMS user responds to this message 'Ok. I'll wait for you at the entrance'. For doing this, the UE simply generates a MESSAGE request, fills in the desired content and populates the Request-URI with the address of the recipient. The message looks like this:

```
MESSAGE tel:+358503333333 SIP/2.0
Via: SIP/2.0/UDP [5555::d:c:b:a]:1357;branch=bj64636
Max-Forwards: 70
Route:<sip:[6666::33:44:55:66]:6701;lr>,sip:orig@scscf1.home1.net;lr
P-Access-Network-Info:3GPP-E-UTRAN-TDD;utran-cell-id-3gpp=2449145353
From: <tel:+358405556667>; tag=34563
To: <tel:+358503333333>
Call-ID: gh54353whwrhw
Cseq: 512 MESSAGE
Require: sec-agree
Proxy-Require: sec-agree
Security-Verify: ipsec-3gpp; alg=hmac-sha-1-96;
spi-c=2345456; spi-s=345346346; port-c=2342; port-s=6701
Content-Type: plain/text
Content-Length: 38
Ok. I'll wait for you at the entrance
```

[15] Here a simplified interworking procedure is shown. If the recipient UE supports Open Mobile Alliance Instant Messaging solution then additional SIP headers (Accept-Contact, User-Agent and Request-Disposition) will be present as defined in 3GPP TS 29.311

The message gets routed to IP-SM-GW at the user's home network which realises that the recipient is not an IMS user and it converts this request to a short message which is sent to the recipient via a SMS centre. The GSM user gets this message as an ordinary SMS and does not see difference from any other SMS.

5.9.3 Multimedia Messaging Service

When a user is attached to LTE there are two alternatives for delivering a MMS notification. The notification is either transporter using SMS over SGs or SMS over IP. This is depicted in Figure 5.43. Once the UE receives the notification it will initiate retrieval of the actual MMS using existing procedures using non IMS APN (e.g. MMS APN). The originating MMS is sent using non IMS APN to the MMS relay/server.

5.9.4 Unstructured Supplementary Services Data Simulation in IMS

For the sake of completeness this chapter also describes the very recently started standardisation topic for 3GPP Release 11 related to support for USSD with IMS-based network architecture. Actual stage 1 level requirement has been already recorded in (3GPP TS 22.173) to support USSD-based procedures to control supplementary services from the serving IMS domain.

Additionally only mobile initiated unstructured supplementary service data (MI-USSD) has been included within the scope of this work. This means that NI-USSD which was introduced as part of USSD phase 2 is excluded with those use cases that has been using that particular functionality today. Therefore complete functional parity cannot be achieved to circuit switched-based USSD as of today.

Additionally the use of USSD has been already possible from LTE attached UE with the use of CS fallback for EPS procedures from 3GPP Release 8 onwards, as described in a previous chapter in this book. However in order to implement the same capability using IMS architecture using the SIP signalling protocol instead of using procedures defined by (3GPP TS 24.008) then this 3GPP Release 11 defined unstructured supplementary services data simulation in IMS (USSI) capability is required.

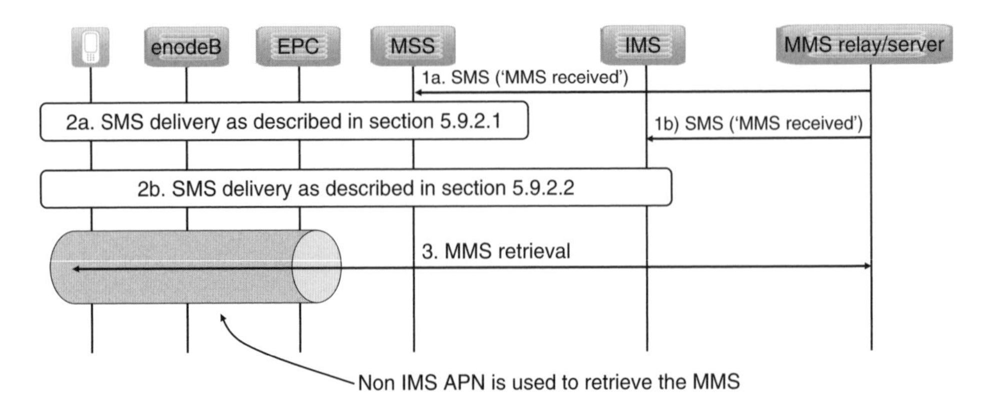

Figure 5.43 Example of terminating MMS.

Due to the non-mature status of work in 3GPP the exact details of the required protocol level changes are not yet known at the time of writing this book. However it is expected that either a SIP message or info request will be used to transfer the USSD payload transparently between the UE and the IMS AS responsible for performing the protocol level translation between the SIP and MAP interfaces. It is also required that a SIP dialogue is established between the UE and AS due the nature of USSD. Eventually this AS can be considered to act similarly to IP-SM-GW in the case of SMS over IP to bridge IMS-based USSD to legacy circuit switched USSD applications.

It is estimated that initial VoLTE deployments will be made without this functionality in place and in the case where USSD capability is required then CS fallback for EPS procedures can be used from a circuit switched network. However it is expected that eventually this functionality will become available as an integrated functionality in LTE UEs.

5.9.5 Summary

As described in this chapter backwards compatibility to a number of different messaging technologies familiar from today's mobile network is considered extremely important when VoLTE services are implemented.

A number of different technologies have been introduced which either re-use circuit switched networks such as SMS over SGs and CS fallback for EPS for USSD use cases but also native IP-based technologies such as SMS over IP and USSI have been disclosed.

As in the case of voice and video telephony it is also foreseen with messaging services that eventually these services will be converted to use native IP-based technologies, but the time it takes to make this transition will vary a lot and it is likely the technologies will be run in parallel in the same network for a long period of time.

6

IMS Centralized Services

IMS Centralized Services (ICS) architecture as specified by 3GPP (3GPP TS 23.292) aims to provide telephony services by using IP multimedia subsystem (IMS)-based service enablers regardless of the used access technology of the end user. IMS-based service enablers relevant to the scope of this book are mainly multimedia telephony services offered by a telephony application server (TAS). However other service enablers that can be offered by IMS architecture can be such as presence services provided by an open mobile alliance (OMA) presence server, various service configuration-related services provided by a XML document management server (XDMS) as well as instant messaging services provided by an OMA instant messaging server to name a few enablers. These service enablers belong within the scope of a rich communication suite (RCS) that is an individual initiative within the global system for mobile communications association (GSMA) to enrich the end user experience.

Additionally all future 3GPP standardisation related to IMS is expected to be based on the existence of ICS-based IMS architecture including features such as multimedia session continuity as defined by 3GPP (3GPP TS 23.237) enabling more flexible control of individual multimedia streams by user equipment (UE) and network as well as single radio voice call continuity (SR-VCC) from UMTS terrestrial radio access network (UTRAN)/GSM/EDGE radio access network (GERAN) to evolved UMTS terrestrial radio access network (E-UTRAN; also called reverse SR-VCC) as defined in (3GPP TS 23.216), to name a few practical examples.

The ICS-based IMS architecture defines principles on how to route both originating and terminating sessions for a long term evolution (LTE) subscriber using services provided by an ICS-based network. In ICS architecture both the originating and terminating calls (regardless of the access technology in use by the served UE) are routed via IMS. This means that originating domain selection by UE as well as terminating domain selection by network need to occur to select the appropriate access domain technology for the session. Additionally this selection process takes into account the requested session description indicating voice, video or other kind of media as well as the access domain capabilities currently existing for the served UE.

Voice over LTE: VoLTE, First Edition. Miikka Poikselkä, Harri Holma, Jukka Hongisto, Juha Kallio and Antti Toskala.

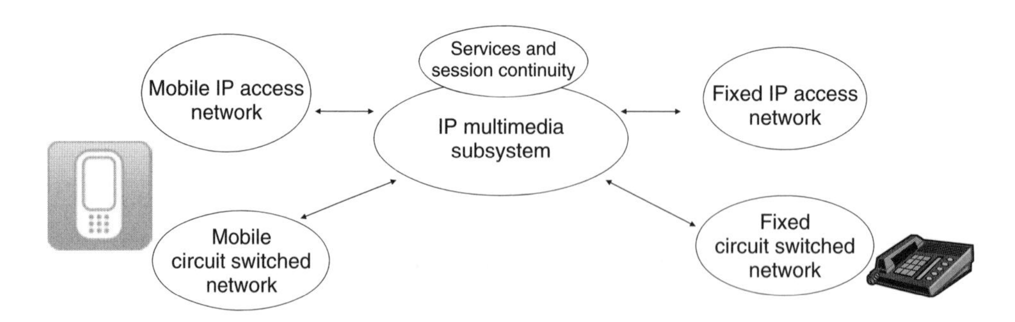

Figure 6.1 High level principle of IMS Centralized Services.

Figure 6.1 represents the high level principle ICS-based IMS architecture providing the same services in access-independent fashion for circuit switched mobile and fixed accesses and also for mobile and fixed broadband accesses. This kind of approach enables the operator to reduce the cost of service creation and deployment by offering the capability to implement a service once for all types of accesses.

Since one of the most essential features of ICS-based IMS architecture is to enable this access-agnostic service execution, the following part of this chapter addresses service consistency related to service execution and end user experience, depending on the type of UE used by the end users.

Service consistency for end user experience related to IMS deployed services depends on the capabilities of the UE as well as the services in question, naturally. One kind of UE that uses a session initiation protocol SIP/Gm-based session control protocol over a packet switched connection to establish sessions may be able still to use a circuit switched access network as a bearer for voice and video connection in the case when an Internet protocol (IP)-based access network cannot provide sufficient quality. Benefit of such approach is that this kind of UE is able to leverage the maximal set of services offered by ICS-based IMS architecture whereas still using traditional CS bearer for voice connection. The second kind of UE uses either circuit switched call control or SIP/Gm-based session control based on the current access network in use; and this kind of UE will always be limited by the capabilities of the currently used access technology. For instance in the case when such a UE is using circuit switched mobile access network, then the service capabilities are limited to those made available via circuit switched call control protocol, namely (3GPP TS 24.008). This limitation is not considered to be an issue with voice over long term evolution (VoLTE) services since in fact VoLTE related multimedia telephony services that are offered by TAS (based on GSMA IR.92) are very similar to the services available today from a circuit switched mobile core network.

The former kind of UE, which is also called an ICS-enhanced UE as defined in (3GPP TS 24.292), maximises the service capabilities that can be offered by network since SIP/Gm signalling is always used as part of session establishment even when the terminal is roaming in a non-voice over Internet protocol (VoIP) capable access network (2G/3G); and in this way it is possible to use alphanumerical uniform resource identifiers (URIs) but also to have enriched content as part of session establishment that could not be present with traditional circuit switched call control protocols (3GPP TS 24.008). In the case of ICS-enhanced UE, if packet switched connectivity for SIP/Gm cannot be established towards the IMS domain for some reason, 3GPP has a defined additional interface that is called I1 to enable communication for session control purposes between UE and serving IMS domain. This I1 interface is transported on top of unstructured supplementary service data (USSD) and defined in (3GPP TS 24.294). It is also a property of the I1 interface to enable switching between native SIP/Gm and I1 in the case when the UE cannot maintain concurrent packet switched connectivity at all times to the IMS domain.

Non-ICS enhanced UEs are sufficient to replicate the end user experience of existing telephony services that can be provided by mobile networks today. This means that the end user is able to have a consistent service experience when roaming either in circuit switched or IP-based access networks without requiring a consistent session control signalling interface towards an ICS enhanced IMS network. It is assumed that in the first phase of VoLTE deployments non-ICS enhanced UEs will be used with ICS enhanced IMS architecture due the fact that GSMA VoLTE profile (IR.92) does not require the more advanced service capabilities that would only be available for ICS enhanced UEs.

Figure 6.2 represents the principal difference between ICS enhanced and non-ICS enhanced UEs at high level. The third kind of UEs that are able to use the services provided by ICS enhanced IMS architecture are native circuit switched UEs which have been sold today in billions. These kind of terminals do not have a native SIP/Gm

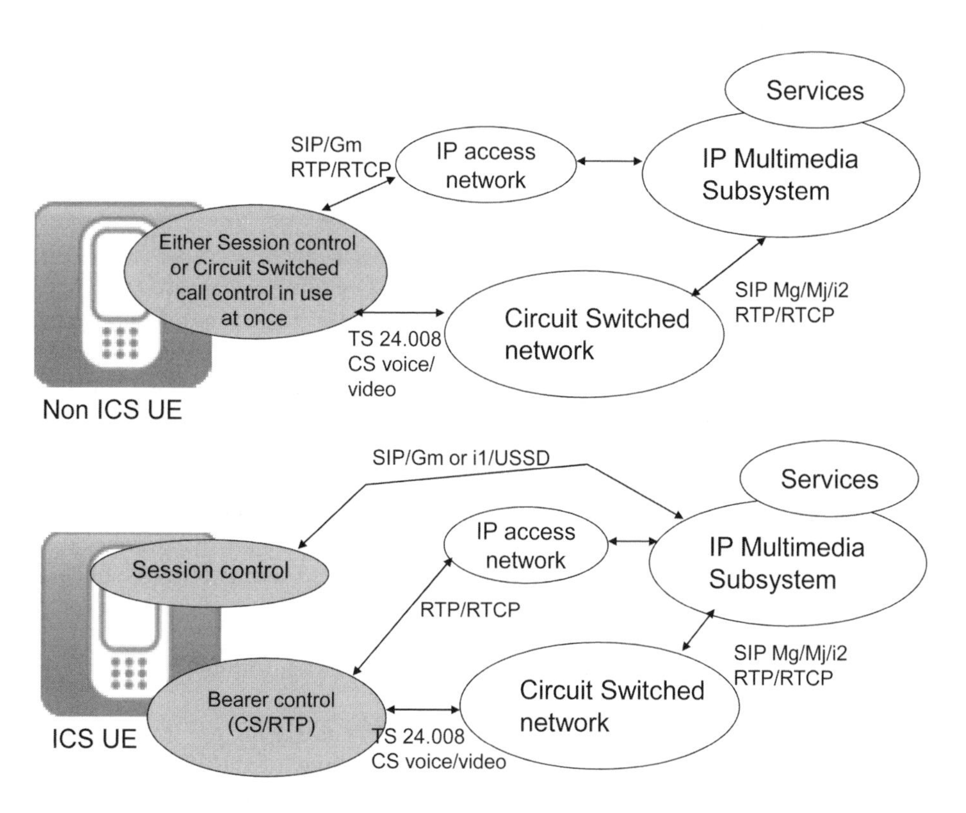

Figure 6.2 Difference between ICS enhanced and non-ICS enhanced UEs.

Table 6.1 Different kinds of UEs and IMS Centralized Service capabilities

Access technology in use	ICS enhanced UE (future UEs)	Non-ICS enhanced UE (VoLTE UEs)	Legacy UE (UEs today)
2G circuit switched access without concurrent use of packet switched data services (dual transfer mode)	CS voice with CS equivalent multimedia telephony services	CS voice with CS equivalent multimedia telephony services	CS voice with CS equivalent multimedia telephony services
3G circuit switched access with concurrent use of packet switched data services (MultiRAB)	CS voice with full multimedia telephony services	CS voice with CS equivalent multimedia telephony services	CS voice with CS equivalent multimedia telephony services
3G packet switched access	IP voice with full multimedia telephony services	IP voice with CS equivalent multimedia telephony services	Not available
LTE packet switched access	IP voice with full multimedia telephony services	IP voice with CS equivalent multimedia telephony services	Not available
Alternative packet switched accesses (WLAN, WiMAX)	IP voice with full multimedia telephony services	IP voice with CS equivalent multimedia telephony services	Not available

WLAN = wireless local area network

interface nor support for VoIP. In such a case the service consistency naturally is the lowest, compared to the previously mentioned UE categories.

In order to illustrate the service consistency, Table 6.1 describes service capabilities when using the different kinds of UEs described above.

Full multimedia telephony services include the capability to use other kinds of media (for instance chat) as part of the same IMS session in addition to voice or video. Also service functionalities provided by the core network would be richer than the services available today in circuit switched networks.

Circuit switched equivalent multimedia telephony services are a subset of full multimedia telephony services and are equivalent to the service set that is available today in mobile circuit switched networks.

For more information about multimedia telephony service capabilities as well as related network architecture, see Section 4.6.

3GPP has also defined a number of architectural enhancements for IMS architecture in order to support ICS. One part of the key functionalities of the ICS architecture is the service centralisation and service continuity application server function (SCC AS), which is responsible for keeping track of all sessions inside IMS that may be targets for domain transfer or session transfer procedures but also to perform actual domain and session transfer procedures when so required. More information about the functionalities provided by SCC AS can be found in Section 5.6.

Another architectural enhancement defined as part of the ICS architecture is named the ICS enhanced mobile switching centre (MSC) server, which is an extension of the MSC server functionality originally defined as part of 3GPP Release 4 to provide services for 2G/3G circuit switched UEs. The ICS enhanced MSC server is able to perform the IMS registration on behalf of non-IMS capable UEs to the serving ICS architecture; and this ensures that ICS service enablers such as TAS or SCC AS are aware of the UE reachability whereas in an architecture having no such functionality such knowledge would not be available.

The use of an ICS enhanced MSC server also optimises both originating and terminating call routing since a call can be immediately routed by using IMS-based routing principles without the need to use, for example intelligent network-based homing functionality to forcedly route call via IMS, as typically would be required today in case ICS architecture is used. Additionally the ICS enhanced MSC server is defined to provide the capability to convert a supplementary service related service management received from UE as described in (3GPP TS 24.010) to a XML configuration access protocol (XCAP)-based supplementary service control protocol as described in (3GPP TS 24.623) and (3GPP TS 24.173).

Figure 6.3 represents ICS architecture using ICS enhanced MSC server functionality to enable legacy circuit switched UEs to access ICS services.

A careful reader is able to see from Figure 6.3 that the same MSC server which is ICS enhanced is also enhanced with SR-VCC capability, as described in Section 5.6.

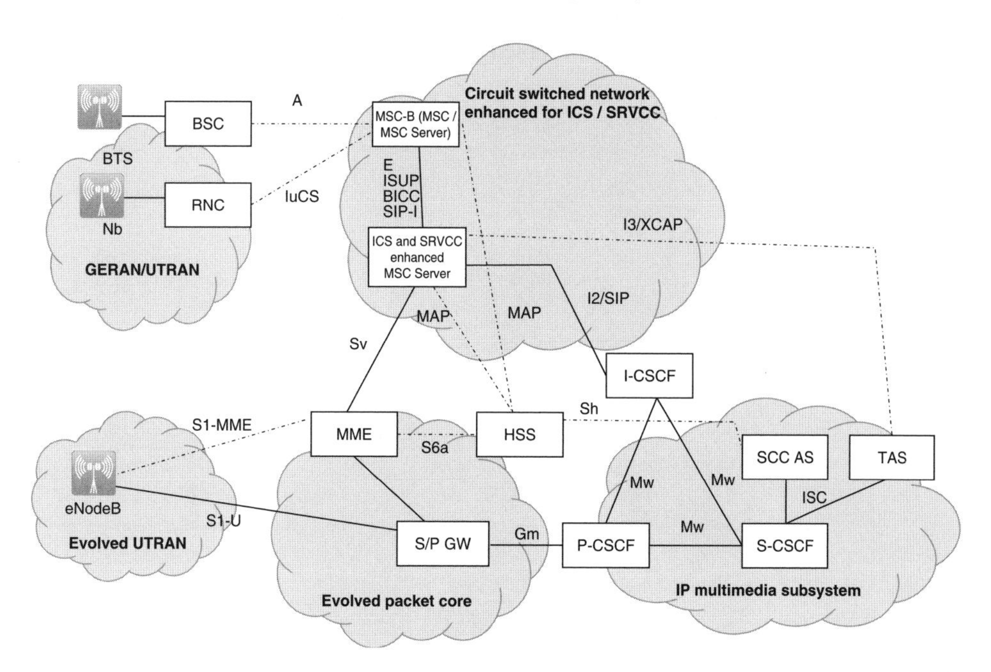

Figure 6.3 IMS Centralized Service architecture.

This drawing illustrates the evolution of SR-VCC towards what has been standardised from 3GPP Release 9 onwards to provide capability at network level to support domain transfer for mid-call services. In such situation the SR-VCC enhanced MSC server also at same time acts similarly to ICS enhanced MSC server and uses the i2-based SIP interface towards the serving IMS domain. For more information about the phasing of SR-VCC, see Section 5.6.

As described in this chapter ICS enhanced IMS architecture will enable the operator to deploy, for instance, telephony services in a centralised fashion and to reuse these capabilities in an access-independent fashion from different kind of accesses. ICS enhanced IMS architecture has been taken as a basis for VoLTE but, since the multimedia telephony services are profiled to be aligned with circuit switched services today, the ICS enhanced UEs behaviour is not mandated by VoLTE. This means in practice that UE is able, for instance, to roam outside the IMS domain and to use the currently available circuit switched call control procedures to reach IMS-based services.

This chapter also introduced the capability to support services offered by ICS enhanced IMS architecture for legacy UEs that do not have any SIP/Gm capabilities or native VoIP support. This approach is similar to what was standardised by the European Telecommunications Standards Institute (ETSI) telecommunications and Internet converged services and protocols for advanced networking (TISPAN) for fixed circuit switched access to IMS by using a specific fixed access gateway control function (AGCF). In the case of mobile circuit switched access the functionality is named ICS enhanced MSC server instead.

Finally the use of ICS enhanced IMS architecture enables the operator to introduce new services which are not able to be provided by traditional circuit switched networks for UEs that are ICS enhanced. When and what are those services in detail is naturally the subject for innovations enabled by IP-based technologies such as SIP servlets and other associated development technologies enabled by IMS architecture.

7

VoLTE Radio Performance

This section considers the radio coverage, radio capacity and end to end latency which are key factors for end user voice call quality and for operator network dimensioning.

7.1 Coverage

Long term evolution (LTE) radio has been designed for low latency and high data rates, but providing good coverage requires further optimisation steps. Adaptive multirate (AMR) voice packets arrive every 20 ms and can be transmitted in 1 ms transmission time intervals (TTIs) which allows the terminal power amplifier to transmit only for a short time making uplink coverage a problem. The optimal solution for coverage would allow the terminal to have continuous transmission. The solution in LTE is to use TTI bundling where the same data is repeated in four consecutive TTIs. The other solution is to use layer 1 retransmissions which also increases the effective transmission time. Combining these two solutions also makes LTE coverage competitive for the voice service. TTI bundling and retransmissions are illustrated in Figure 7.1.

TTI bundling requires terminal support. If TTI bundling is not supported, another option is to use medium access control (MAC) segmentation which does not need any specific terminal support. This solution splits the voice packets into smaller ones which again forces the scheduler to use multiple TTIs for voice packet transmission.

The resulting evolved eNodeB sensitivity values with retransmission and TTI bundling are shown in Table 7.1. The eNodeB sensitivity can be calculated as follows:

$$eNodeB\_sensitivity(\text{dBm}) = -174 + 10 \cdot \log_{10}(Bandwidth) + Noise\_figure + SNR$$

The eNodeB receiver noise figure is assumed to be 2 dB, two resource blocks are used for voice and no interference margin is included. The bundling can improve the uplink voice coverage by 4 dB.

The wideband code division multiple access (WCDMA) eNodeB sensitivity can be estimated by assuming Eb/N0 = 4.5 dB, which gives −126.6 dBm. In order to get similar voice coverage in LTE as in WCDMA we need to apply TTI bundling with a sufficient number of retransmissions.

Voice over LTE: VoLTE, First Edition. Miikka Poikselkä, Harri Holma, Jukka Hongisto, Juha Kallio and Antti Toskala.

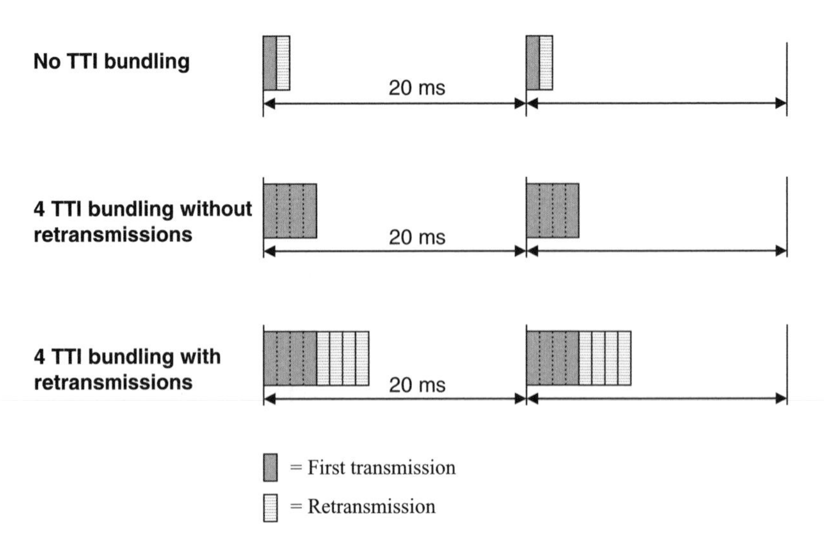

Figure 7.1 TTI bundling and retransmissions for improving uplink coverage. Reproduced with permission from John Wiley & Sons, Ltd.

Table 7.1 Uplink VoIP sensitivity with TTI bundling

Number of TTIs bundled	One	Four
Transmission bandwidth	360 kHz (two resource blocks)	360 kHz (two resource blocks)
Number of retransmissions	Six	Three
Required SNR	–4.2 dB	–8.0 dB
Receiver sensitivity	–120.6 dBm	–124.4 dBm

7.2 Capacity

Voice capacity simulations are illustrated for downlink in Figure 7.2 and for uplink in Figure 7.3. These simulations assume three-sector macro cell deployment with voice only traffic. Three different cases are shown: dynamic scheduling without packet bundling, dynamic scheduling with packet bundling and semi-persistent scheduling. Three different AMR voice codec data rates are studied: 12.2, 7.95 and 5.9 kbps. The lower data rates give higher capacity with packet bundling and with semi-persistent scheduling. If dynamic scheduling is used without packet bundling, the capacity does not increase with lower AMR data rate because the capacity is limited by the control channels. The maximum uplink capacity is more than 200 simultaneous voice users per 5 MHz per sector both in uplink and in downlink with AMR 12.2 kbps. The capacity would be similar with wideband AMR 12.65 kbps. The lower rate AMR 5.9 kbps allows even more than 400 users per 5 MHz.

These capacity simulations assume that Internet protocol (IP) header compression [Robust header compression (RoHC)] is used. Header compression is essential for voice

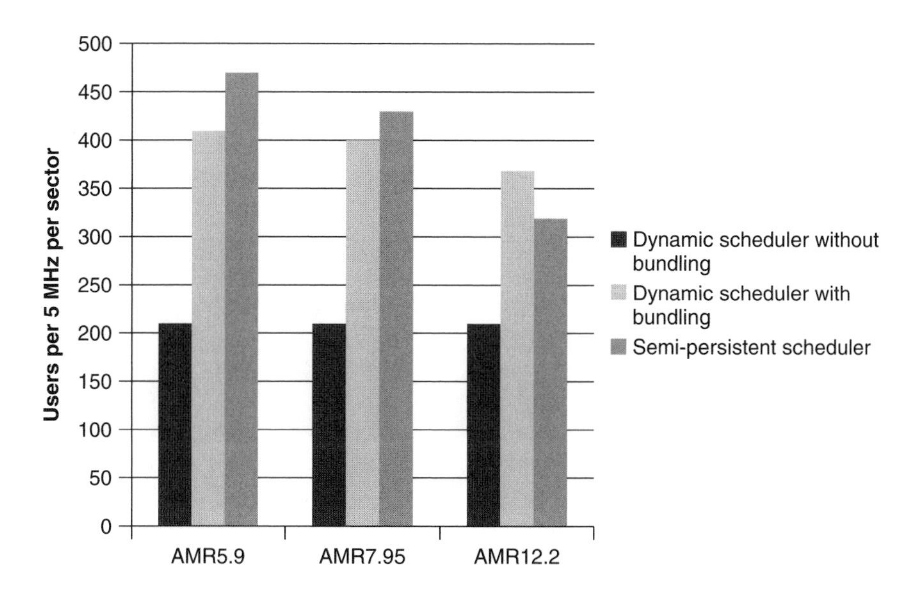

Figure 7.2 Downlink voice capacity per 5 MHz per sector.

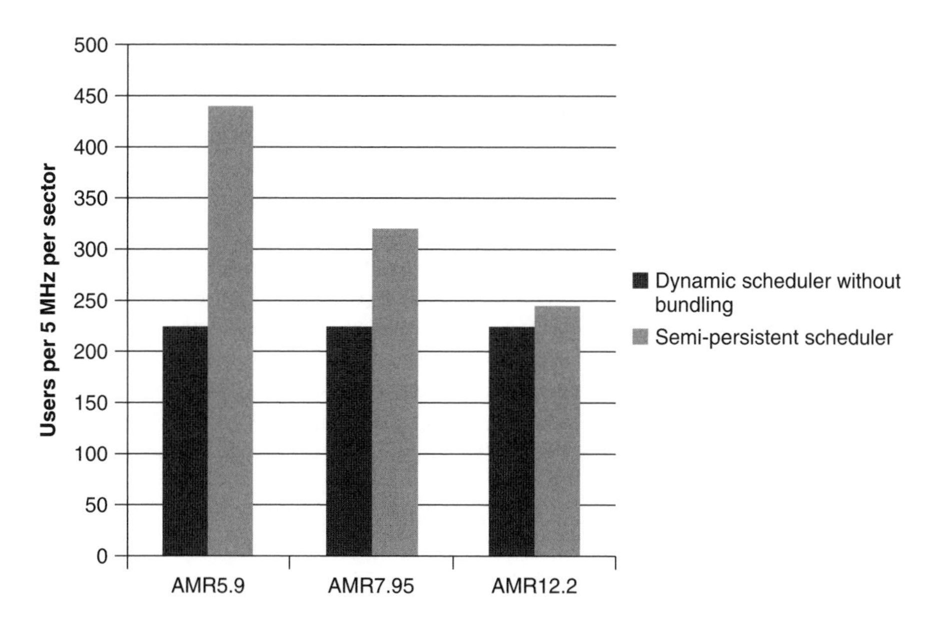

Figure 7.3 Uplink voice capacity per 5 MHz per sector.

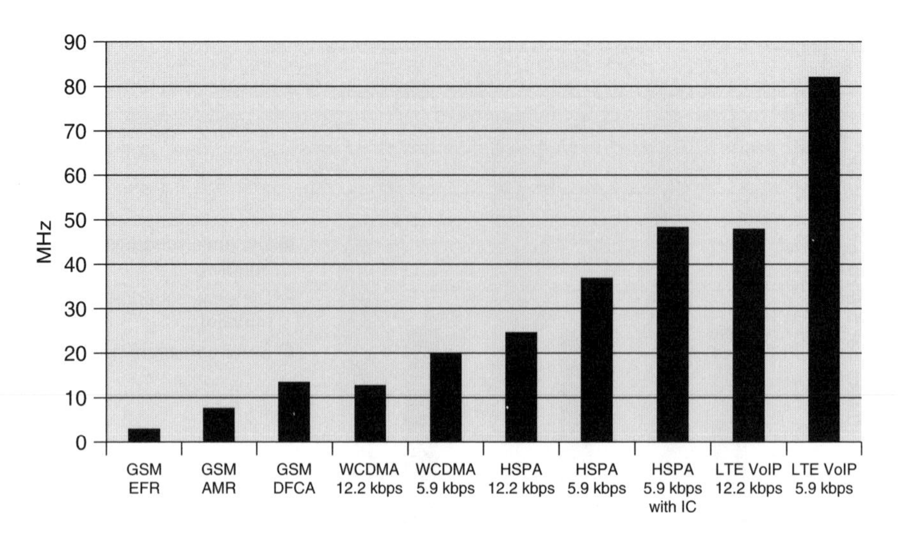

Figure 7.4 Evolution of voice spectral efficiency. Reproduced with permission from John Wiley & Sons, Ltd.

over Internet protocol (VoIP) to provide the maximum capacity. If full IP headers are transmitted, then more than 50% of the transmission is wasted on transmitted IP headers because the headers are even larger than the size of the 12 kbps voice packet. The header compression is done in eNodeB and in user equipment (UE).

The evolution of voice spectral efficiency is illustrated in Figure 7.4. The efficiency is shown as maximum simultaneous voice users per megahertz per sector in a macro cellular network. The impact of frequency reuse is included in the global system for mobile communications (GSM) calculation. The early GSM with a fixed voice codec rate [enhanced full rate (EFR)] gives 4 users/MHz. The latest GSM features can triple the GSM efficiency to beyond 10 users/MHz by using AMR codec adaptation, dynamic frequency and channel allocation (DFCA) and terminal single antenna interference cancellation (SAIC). Third generation high speed packet access (3G HSPA) voice solutions with AMR 5.9 kbps increases the capacity beyond 40 users/MHz. LTE further doubles the capacity to 80 users – up to 20 times more than in the early GSM. The high LTE voice efficiency allows carrying the voice traffic in smaller spectrum, which leaves more room for the increasing data traffic.

The high efficiency allows squeezing the voice traffic into a smaller spectrum. An example calculation is shown in Figure 7.5 assuming 1500 subscribers per sector and 40 mErl traffic per subscriber. GSM EFR would take 25 MHz of spectrum while LTE will be able to carry that voice traffic in less than 2 MHz. LTE can then free up more spectrum for data usage.

7.3 Latency

The end to end delay budget for LTE VoIP is considered here. The delay should preferably be below 200 ms which is the value typically achieved in circuit switched networks

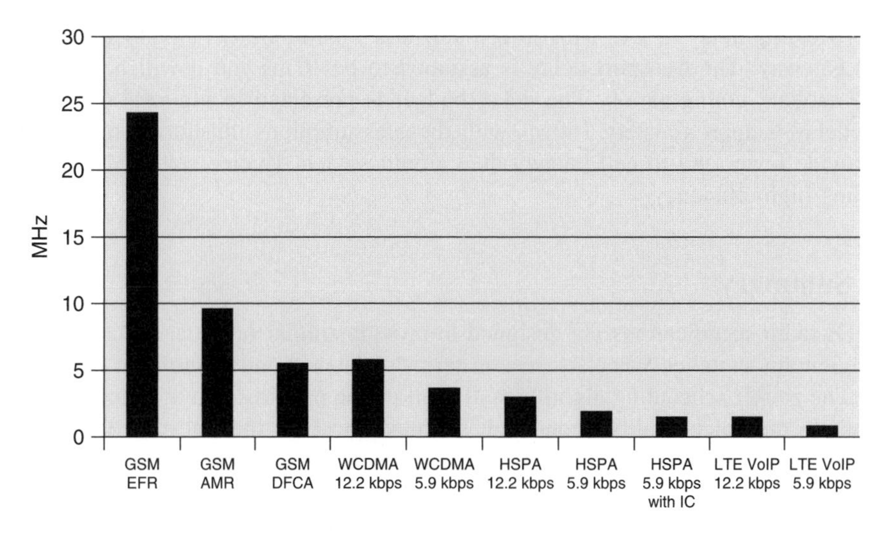

Figure 7.5 Spectrum required for supporting 1500 subscribers per sector at 40 mErl. Reproduced with permission from John Wiley & Sons, Ltd.

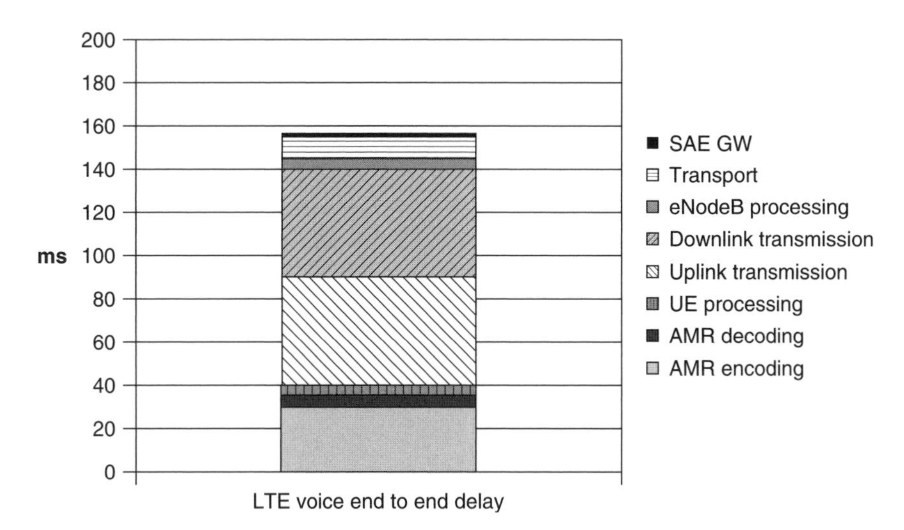

Figure 7.6 LTE voice end to end delay budget. Reproduced with permission from John Wiley & Sons, Ltd.

today. We use the following assumptions in the delay budget calculations. The voice encoding delay is assumed to be 30 ms, including 20 ms frame size, 5 ms look ahead and 5 ms processing time. The receiving end assumes 5 ms processing time for the decoding. The capacity simulations assume maximum 50 ms air–interface delay in both uplink and downlink including scheduling delay and the time required for the initial and the hybrid automatic repeat request (HARQ) retransmissions of a packet. We also assume

5 ms processing delay in UE, 5 ms in eNodeB and 1 ms in system architecture evolution (SAE) gateway. The transport delay is assumed to be 10 ms and it will heavily depend on the network configuration. The delay budget is presented in Figure 7.6. The mouth to ear delay is approximately 160 ms with these assumptions illustrating that LTE VoIP can provide lower end to end latency than circuit switched voice calls today while still providing high efficiency.

7.4 Summary

The LTE radio specifications are designed to provide similar or better voice quality than existing circuit switched voice services in terms of latency and in terms of network coverage. The packet scheduling algorithms also allow the prioritisation of voice connections to maintain the voice quality under high loading. The LTE spectral efficiency for voice is clearly higher than in 2G or in 3G radio systems, which allows LTE to carry good quality voice in less spectrum and to use more spectrum for the growing data traffic with excellent spectrum efficiency as studied in (Holma, H. and Toskala, A. "LTE for UMTS", 2nd edition, 2011).

8

HSPA Voice over IP

In the previous chapters we described how voice over long term evolution (VoLTE) works and we believe VoLTE is a driver for mobile voice over Internet protocol (VoIP) and a technology where mobile VoIP will take off first. However, we also believe that voice over high speed packet access (VoHSPA) could be used to speed up VoIP deployment by having a nationwide IP network with a combination of two technologies in a fast and cost-effective way (no need to have LTE everywhere to provide unified VoIP experience). When VoIP is deployed in HSPA as well, then a packet (see Section 5.6.1) handover can be used from LTE to HSPA instead of single radio voice call continuity (SR-VCC) to minimise impact to legacy circuit switched (CS) networks. Also VoHSPA boosts the (wideband code division multiple access) WCDMA/HSPA network's voice capacity (see Section 7.2) compared to the current CS voice deployments and it further gives the possibility to start re-farming voice from CS [mainly from global system for mobile communications (GSM)] to IP (LTE/HSPA) and makes it possible for operators to release some GSM frequencies to LTE.

The very same IP multimedia subsystem (IMS) functionality which is required for VoLTE (as described in Chapter 5) can be used to provide VoHSPA. The efficient deployment of VoHSPA requires the following essential radio and packet core capabilities:

- Support of the Robust header compression (RoHC). Since the IP header itself without compression is large compared to the voice payload itself, one needs to use RoHC in order not to degrade the system voice capacity. With Internet protocol version 4 (IPv4) already the uncompressed IP header would drop the voice capacity to half, with an even larger header size in IPv6.
- Use of high-speed downlink packet access (HSDPA) and high speed uplink packet access (HSUPA). The VoIP can also be an operator on top of Release 99 WCDMA, but the capacity and power consumption benefits need the use of HSDPA and HSUPA. In order to obtain the full potential one should run both the VoIP as well as the necessary signalling traffic over HSDPA/HSUPA to reach a good capacity. Running all the traffic on HSDPA (and HSUPA) minimises the overhead because a continuous dedicated traffic channel (DCH) does not need to be maintained between voice packets.

Voice over LTE: VoLTE, First Edition. Miikka Poikselkä, Harri Holma, Jukka Hongisto, Juha Kallio and Antti Toskala.

- Release 7 onwards from HSPA evolution had added features to the HSPA specifications, especially continuous packet connectivity (CPC), which also allows the user equipment (UE) in Cell_DCH state to have discontinuous reception and transmission. This is beneficial for a voice service where packets arrive at regular 20 ms intervals. This enables a great improvement on the talk time over the Release 99-based voice implementation which requires continuous reception and transmission. The CPC principle is illustrated in Figure 8.1. Another dimension of the development is the equaliser receiver for HSDPA channels, which allows an improvement in the downlink performance when compared to the Release 99 DCH. With VoIP the use of HS-SCCH-less operation reduces the HSDPA signalling overhead, benefitting from the constant packet inter-arrival time.
- The network features which can further improve the VoHSPA capacity include packet bundling and VoIP optimised scheduler. The latter is enabled by ensuring that VoHSPA packets are not handled as background data but rather with the proper quality of service (QoS) class as well as from core network but also from the radio access network (RAN) point of view as well. The VoHSPA with signalling needs to be mapped on such PS radio access bearers (RABs) that ensure proper handling at the scheduler (PS conversational RAB, PS Interactive RABs).
- The use of VoHSPA allows not only higher capacity over Release 99-based CS, as shown in Figure 8.2, but also there are big benefits from the terminal talk time point of view. As presented in Figure 8.2, there is an estimated 80% improvement in the talk time, taking the 2011 technology components as the analysis basis when comparing the power consumption of voice service on Release 99 DCH and running voice on HSPA with power saving features (mainly CPC) enabled. Such benefits could be achieved also with CS voice by using the mapping as specified in Release 8 which enables mapping the CS voice also on HSPA, as explained in (Holma, H. and Toskala, A. "WCDMA for UMTS", 5th edition, 2010).

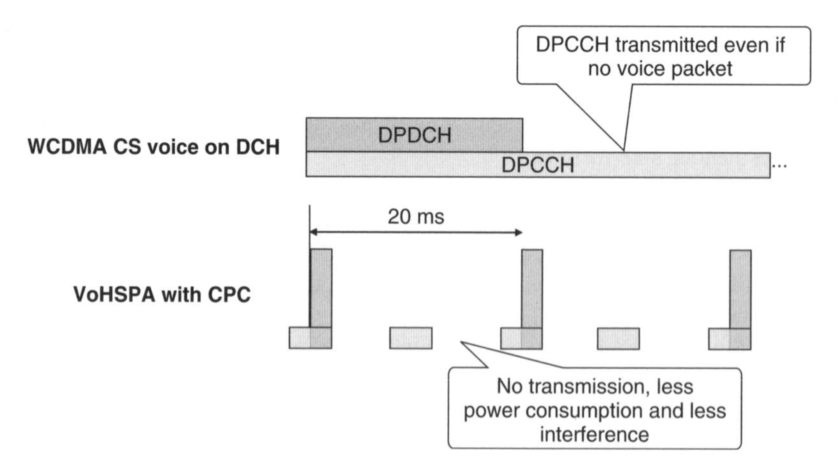

Figure 8.1 HSPA continuous packet connectivity (CPC) principle (DPDCH = dedicated physical data channel; DPCCH = dedicated physical control channel).

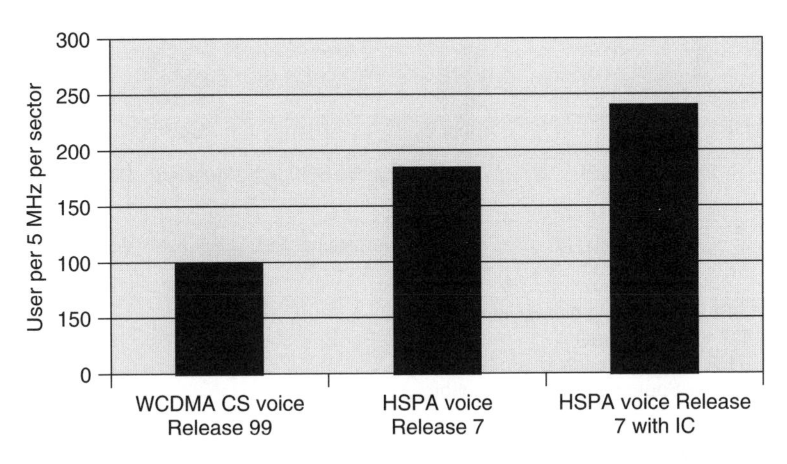

Figure 8.2 Capacity benefit from VoHSPA when compared to CS voice on WCDMA DCH (IC = interference cancellation).

LTE	HSPA				
QCI	Traffic class	Traffic handling priority	Signalling indicator	Source statistics descriptor	
1	Conversational	-	-	Speech	Voice
2	Conversational	-	-	-	
3	Conversational	-	-	-	
4	Streaming	-	-	-	
5	Interactive	1	Yes	-	IMS
6	Interactive	1	No	-	
7	Interactive	2	No	-	
8	Interactive	3	No	-	
9	Background	-	-	-	

Figure 8.3 Mapping of LTE and HSPA QoS parameters for VoIP service.

The interface between the core network and radio HSPA architecture is based on the Iu interface compared to the S1 in LTE. While the radio parameterisation was visible to the core network, the correct QoS parameters were important. In LTE the approach was based on the quality class identifier (QCI) use while with the HSPA network there are specific QoS parameters in use, as shown in Figure 8.3.

The serving GPRS support node (SGSN) provides over the Iu interface to the RNC information on the bearer level including QoS information, as presented in Figure 8.3. In the case of VoHSPA the traffic class to be used is *interactive* with *THP* = 1 for the

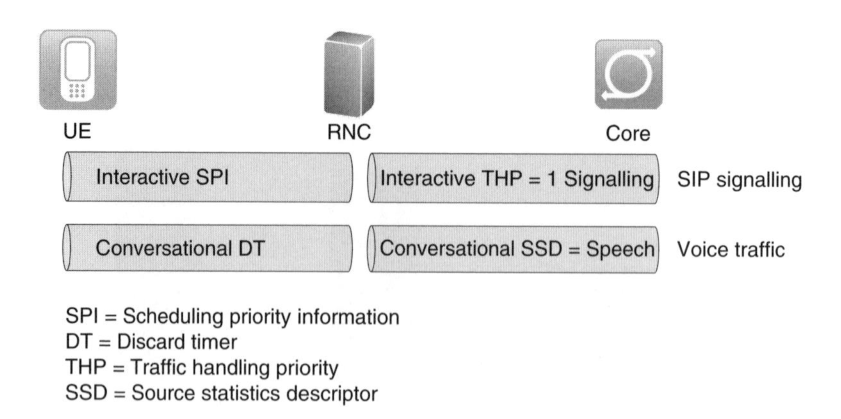

Figure 8.4 QoS parameters from packet core to HSPA RAN.

IMS signalling with maximum bit rate, and *conversational* for the voice media with a guaranteed bit rate. The RNC then converts these towards the NodeB over the Iub interface (unless the RNC functionality is co-located with NodeB in flat HSPA architecture) to the scheduling priority indicator, discard time to be used as well as guaranteed bit information. The HSPA scheduler then takes these parameters into account to ensure appropriate scheduler behaviour to the voice traffic on HSPA. The parameter mapping is illustrated in Figure 8.4.

As a summary from VoHSPA the benefit from also running VoHSPA is both on the capacity side while also at the same time reducing the complexity and time needed for inter-system handover when one can stay within a PS domain when moving between HSPA and LTE networks with an active voice connection. The recent developments on HSPA also enable one to move to a new level in voice power consumption with the use of HSPA power saving features to run VoHSPA while at the same time boosting the capacity with a reduced control signal overhead and other improvements in HSPA technology.

References

IR.65 IMS Roaming and Interworking Guidelines.
IR.92 IMS Profile for Voice and SMS.
GSMA http://www.gsmworld.com/newsroom/press-releases/2010/4634.htm (accessed January 21 2011).
RFC2401 Security Architecture for the Internet Protocol.
RFC2486 The Network Access Identifier.
RFC2617 HTTP Authentication: Basic and Digest Access Authentication.
RFC2833 RTP Payload for DTMF Digits, Telephony Tones and Telephony Signals.
RFC3261 SIP: Session Initiation Protocol.
RFC3262 Reliability of Provisional Responses in the Session Initiation Protocol (SIP).
RFC3264 An Offer/Answer Model with SDP.
RFC3265 Session Initiation Protocol (SIP)-Specific Event Notification.
RFC3310 Hypertext Transfer Protocol (HTTP) Digest Authentication Using Authentication and Key Agreement (AKA).
RFC3311 The Session Initiation Protocol (SIP) UPDATE Method.
RFC3312 Integration of Resource Management and Session Initiation Protocol.
RFC3323 A Privacy Mechanism for the Session Initiation Protocol (SIP).
RFC3325 Private Extensions to the Session Initiation Protocol (SIP) for Asserted Identity within Trusted Networks.
RFC3327 Session Initiation Protocol (SIP) Extension Header Field for Registering NonAdjacent Contacts.
RFC3329 Security Mechanism Agreement for the Session Initiation Protocol (SIP).
RFC3455 Private Header (P-Header) Extensions to the Session Initiation Protocol (SIP) for the 3rd-Generation Partnership Project (3GPP).
RFC3550 RTP: A Transport Protocol for Real-time Applications.
RFC3551 RTP Profile for Audio and Video Conferences with Minimal Control.
RFC3556 SDP Bandwidth Modifiers for RTCP Bandwidth.
RFC3608 Session Initiation Protocol (SIP) Extension Header Field for Service Route Discovery during Registration.
RFC3680 A Session Initiation Protocol (SIP) Event Package for Registrations.
RFC3840 Indicating User Agent Capabilities in the Session Initiation Protocol (SIP).
RFC3841 Caller Preferences for the Session Initiation Protocol (SIP).
RFC3842 A Message Summary and Message Waiting Indication Event Package for the Session Initiation Protocol (SIP).
RFC3986 Uniform Resource Identifier (URI): Generic Syntax.
RFC4006 Diameter Credit-Control Application.
RFC4566 SDP: Session Description Protocol.
RFC4585 Extended RTP Profile for Real-time Transport Control Protocol (RTCP)-Based Feedback (RTP/AVPF).

Voice over LTE: VoLTE, First Edition. Miikka Poikselkä, Harri Holma, Jukka Hongisto, Juha Kallio and Antti Toskala.

RFC4596 Guidelines for Usage of the Session Initiation Protocol (SIP) Caller Preferences Extension.
RFC4745 Common Policy: A Document Format for Expressing Privacy Preferences.
RFC4867 RTP Payload Format and File Storage Format for the Adaptive Multi-Rate (AMR) and Adaptive Multi-Rate Wideband (AMR-WB) Audio Codecs.
RFC4975 The Message Session Relay Protocol (MSRP).
RFC5031 A Uniform Resource Name (URN) for Emergency and Other Well-Known Services.
RFC5279 A Uniform Resource Name (URN) Namespace for the 3rd Generation Partnership Project (3GPP).
RFC6050 A Session Initiation Protocol (SIP) Extension for the Identification of Services.
3GPP TR 23.815 Charging Implications of IMS Architecture.
3GPP TR 23.882 3GPP System Architecture Evolution (SAE): Report on Technical Options and Conclusions.
3GPP TR 23.893 Feasibility Study on Multimedia Session Continuity.
3GPP TS 22.101 Service Principles.
3GPP TS 22.173 IP Multimedia Core Network Subsystem (IMS) Multimedia Telephony Service and Supplementary Services.
3GPP TS 22.228 Service Requirements for the IP Multimedia Core Network Subsystem.
3GPP TS 23.002 Network Architecture.
3GPP TS 23.003 Technical Specification Group Core Network; Numbering, Addressing and Identification.
3GPP TS 23.009 Handover Procedures.
3GPP TS 23.018 Basic Call Handling; Technical Realization.
3GPP TS 23.040 Technical Realization of the Short Message Service (SMS).
3GPP TS 23.167 IP Multimedia Subsystem (IMS) Emergency Sessions.
3GPP TS 23.203 Policy and Charging Control Architecture.
3GPP TS 23.204 Support of Short Message Service (SMS) Over Generic 3GPP Internet Protocol (IP) Access.
3GPP TS 23.207 End-to-End QoS Concept and Architecture.
3GPP TS 23.221 Architectural Requirements.
3GPP TS 23.228 IP Multimedia (IM) Subsystem; Stage 2.
3GPP TS 23.236 Intra-domain Connection of Radio Access Network (RAN) Nodes to Multiple Core Network (CN) Nodes.
3GPP TS 23.237 IP Multimedia Subsystem (IMS) Service Continuity.
3GPP TS 23.272 Circuit Switched (CS) Fallback in Evolved Packet System (EPS).
3GPP TS 23.292 IP Multimedia Subsystem (IMS) Centralized Services.
3GPP TS 23.334 IP Multimedia Subsystem (IMS) Application Level Gateway (IMS-ALG) – IMS Access Gateway (IMS-AGW) Interface: Procedures Descriptions.
3GPP TS 23.401 General Packet Radio Service (GPRS) Enhancements for Evolved Universal Terrestrial Radio Access Network (E-UTRAN) Access.
3GPP TS 23.402 Architecture Enhancements for non-3GPP Accesses.
3GPP TS 24.008 Mobile Radio Interface Layer 3 Specification; Core Network Protocols.
3GPP TS 24.010 Mobile Radio Interface Layer 3; Supplementary Services Specification; General Aspects.
3GPP TS 24.011 Point-to-Point (PP) Short Message Service (SMS) Support on Mobile Radio Interface.
3GPP TS 24.173 IMS Multimedia Telephony Service and Supplementary Services; Stage 3.
3GPP TS 24.229 IP Multimedia Call Control Based on SIP and SDP; Stage 3.
3GPP TS 24.292 IP Multimedia (IM) Core Network (CN) Subsystem Centralized Services (ICS); Stage 3.
3GPP TS 24.294 IP Multimedia Subsystem (IMS) Centralized Services (ICS) Protocol Via I1 Interface.
3GPP TS 24.301 Non-Access-Stratum (NAS) Protocol for Evolved Packet System (EPS); Stage 3.
3GPP TS 24.341 Support of SMS Over IP Networks; Stage 3.
3GPP TS 24.623 Extensible Markup Language (XML) Configuration Access Protocol (XCAP) Over the Ut interface for Manipulating Supplementary Services.
3GPP TS 24.930 Signalling Flows for the Session Setup in the IP Multimedia Core Network Subsystem (IMS) Based on Session Initiation Protocol (SIP) and Session Description Protocol (SDP); Stage 3.
3GPP TS 25.413 UTRAN Iu interface Radio Access Network Application Part (RANAP) Signalling.
3GPP TS 26.114 IP Multimedia Subsystem (IMS); Multimedia Telephony; Media Handling and Interaction.
3GPP TS 29.002 Mobile Application Part (MAP) Specification.
3GPP TS 29.018 General Packet Radio Service (GPRS); Serving GPRS Support Node (SGSN) – Visitors Location Register (VLR); Gs Interface Layer 3 Specification.
3GPP TS 29.118 Mobility Management Entity (MME) – Visitor Location Register (VLR) SGs Interface Specification.

3GPP TS 29.163 Interworking Between the IP Multimedia (IM) Core Network (CN) Subsystem and Circuit Switched (CS) Networks.
3GPP TS 29.212 Policy and Charging Control Over Gx Reference Point.
3GPP TS 29.213 Policy and Charging Control Signalling Flows and Quality of Service (QoS) Parameter Mapping.
3GPP TS 29.214 Policy and Charging Control Over Rx Reference Point.
3GPP TS 29.228 IP Multimedia (IM) Subsystem Cx and Dx Interfaces; Signaling Flows and Message Contents.
3GPP TS 29.229 Cx and Dx Interfaces Based on the Diameter Protocol; Protocol Details.
3GPP TS 29.272 Evolved Packet System (EPS); Mobility Management Entity (MME) and Serving GPRS Support Node (SGSN) Related Interfaces Based on Diameter Protocol.
3GPP TS 29.280 Evolved Packet System (EPS); 3GPP Sv Interface (MME to MSC, and SGSN to MSC) for SRVCC.
3GPP TS 29.329 Sh Interface Based on the Diameter Protocol.
3GPP TS 29.334 IMS Application Level Gateway (IMS-ALG) – IMS Access Gateway (IMS-AGW); Iq Interface; Stage 3.
3GPP TS 32.240 Telecommunication Management; Charging Management; Charging Architecture and Principles.
3GPP TS 32.260 Telecommunication Management; Charging Management; IP Multimedia Subsystem (IMS) Charging.
3GPP TS 32.295 Charging Management; Charging Data Record (CDR) Transfer.
3GPP TS 32.299 Telecommunication Management; Charging Management; Diameter Charging Applications.
3GPP TS 33.102 3G Security; Security Architecture.
3GPP TS 33.203 3G Security; Access Security for IP-based Services.
3GPP TS 33.210 3G Security; Network Domain Security (NDS); IP Network Layer Security.
3GPP TS 33.220 3G Security; Generic Authentication Architecture (GAA); Generic Bootstrapping Architecture, 3GPP.
3GPP TS 33.401 3GPP System Architecture Evolution (SAE); Security Architecture.
3GPP TS 36.306 User Equipment (UE) Radio Access Capabilities.
3GPP TS 36.331 Radio Resource Control (RRC); Protocol Specification.
3GPP TS 36.413 Evolved Universal Terrestrial Radio Access Network (E-UTRAN); S1 Application Protocol (S1AP).
3GPP TS 48.008 Mobile Switching Centre – Base Station System (MSC-BSS) Interface; Layer 3 Specification.
Holma, H. and Toskala, A. (eds) (2010) *WCDMA for UMTS*, 5th edn, John Wiley & Sons, Inc., New York.
NGMN http://www.ngmn.org/news/partnernews/newssingle0/article/ngmn-alliance-delivers-operatorss-agreement-to-ensure-roaming-for-voice-over-lte-357.html?tx_ttnews[backPid]=3&cHash=7e3a47d38d (accessed January 21 2011).

Index

Voice over LTE: VoLTE, First Edition. Miikka Poikselkä, Harri Holma, Jukka Hongisto, Juha Kallio and Antti Toskala.